Geology and Our National Parks

Other books by Patrick Nurre: These are all also available with sample rock, mineral, and fossil kits at NorthwestRockAndFossil.com.

Rocks and Minerals for Little Eyes (PreK-3)
Fossils and Dinosaurs for Little Eyes (PreK-3)
Volcanoes for Little Eyes (PreK-3)
Geology for Kids (and Geology Journal) (3-6)
Rock Identification Made Easy (3-12)
Rock Identification Field Guide
Fossil Identification Made Easy (3-12)
Fossil Identification Field Guide
Bedrock Geology (high school)
Rocks and Minerals: The Stuff of the Earth (high school)
Volcanoes, Volcanic Rocks and Earthquakes (high school)
The Geology of Yellowstone – A Biblical Guide
Genesis Rock Solid – A Biblical View of Geology
Fossils, Dinosaurs and Cave Men (high school)
Geology and the Hawaiian Islands

Geology and our National Parks
Lessons in Earth Science

Patrick Nurre

The Northwest Treasures Curriculum Project
"Building Faith for a Lifetime of Faith"

Geology and Our National Parks, Lessons in Earth Science
Published by Northwest Treasures
Bothell, Washington
425-488-6848
NorthwestRockAndFossil.com
northwestexpedition@msn.com
Copyright 2018 by Patrick Nurre.
All rights reserved.

Printed in the United States of America. No part of this book may be reproduced in any manner whatsoever without written permission except in the case of brief quotations embodied in critical articles and reviews.

Scripture quotations taken from the New American Standard Bible®.
Copyright © 1960, 1962, 1963, 1968, 1971, 1972, 1973,
1975, 1977, 1995 by The Lockman Foundation
Used by permission. (www.Lockman.org)

Title page photo: Inside cover credit: Canyonlands National Park, Utah By The original uploader was Molas at English Wikipedia - Transferred from en.wikipedia to Commons by Leoboudv using CommonsHelper., CC BY-SA 2.5, https://commons.wikimedia.org/w/index.php?curid=6525027

Contents

Introduction – How to use this book..7
Lesson One – The Framework for Interpreting the Geology of Our National Parks................................8
Lesson Two – Our Restless Earth:
    Yellowstone National Park..14
Lesson Three – Fossils, Preservation and Petrification:
    Petrified Forest National Park..27
    Hagerman Fossil Beds National Monument..35
    John Day Fossil Beds National Monument..40
    Fossil Butte National Monument...46
    Badlands National Park..51
Lesson Four – Stratovolcanoes and Andesite:
    Mt. Rainier National Park..56
    Lassen Volcanic National Park...61
Lesson Five – Glaciers and the Ice Age:
    Yosemite National Park..65
Lesson Six – Orogeny:
    Grand Teton National Park..70
Lesson Seven – Erosion, Slow and Fast:
    The Five National Parks of Utah..74
Lesson Eight – Calderas and Rhyolite:
    Crater Lake National Park..84
    Newberry Volcano/Caldera National Monument..90
Lesson Nine – Basalt Behavior:
    Hawai'i Volcanoes National Park...96
    Craters of the Moon National Monument..105
    Capulin Volcano National Monument..108
    El Malpais National Monument..110
    Lava Beds National Monument..113
Lesson Ten – Global Warming and Climate Change:
    Glacier National Park..117
Lesson Eleven – Deposition and Transportation, The Power of Water:
    Grand Canyon National Park..123
Lesson Twelve – The Rock Types, How Much Have We Observed:
    North Cascades National Park..136
Lesson Thirteen – Interpretive Frameworks and Biases:
    Mount St. Helens National Volcanic Monument..147
Lesson Fourteen – Dinosaurs and The Flood:
    Dinosaur National Monument..158
Lesson Fifteen – Caves and Chemistry:
    Carlsbad Caverns National Park..166
    Mammoth Cave National Park..169
Appendix A – What is Deisim?...173
Appendix B – An Expanded Explanation of Radiometric Dating..176
Appendix C – Final Exam 185
Appendix D – Rock-forming Minerals..189
Appendix E – Answers to Final Exam...190
Index of Words from Word Challenges..191
Credits..194

Introduction
How to use this book

What better way to study Earth Science than to study the geology of God's handiwork through our national parks and monuments? The word *geology* means, *the study of the earth*. It includes both the science *(Earth science)* and the history *(Earth history)* of our Earth.

This book is meant to give the student of geology a Biblical view of Earth history. For over 200 years, civilization has been subject to a secular interpretation of the landforms, rocks and fossils. My goal in this book is to help you develop an interpretation that does not leave the God, who created nature, out of that interpretation. You will be given short Earth science and Earth history lessons that illustrate geological principles through the study of each of the national parks or national monuments selected for this book. The lessons will be reinforced through pictures, word challenges, thought questions, activities and a comprehensive test to ensure that you are developing the critical skills necessary in correctly interpreting Earth science and Earth history. Be sure to acquire a lab book, such as a spiral notebook, to record the definitions of the *Word Challenges*. You will also want to keep track of what you are learning, especially when you have activities to complete. Take your time working through the lessons and thought questions. You will learn the basic principles of geology by doing so. I would suggest that before you begin a particular lesson, that you take the time to look up the *Word Challenges* and record their definitions in your lab book. Some of the definitions will be found in the book, but some will not. You will also find some overlap with certain words in various lessons. This is because each lesson stands alone and doesn't necessarily need to be done in consecutive order.

Many different types of rocks are referenced in this book, along with the minerals that are in them. A chart of the major rock-forming minerals can be found in Appendix D.

This book is meant to be studied with a set of rocks, minerals and fossils that illustrate various geological concepts in our national parks and monuments listed in this book. Although these are not absolutely necessary to completing the study, they do provide a crucial component of study and a perspective that just cannot be acquired by simply looking at pictures. It is highly recommended that you purchase a set of samples either from Northwest Treasures or from some other source.

Upon completion of the course work in this book, the student should have a good working knowledge of both the Earth science and the Earth history, and a practical guide in how to interpret the wonders of nature, in our national parks and national monuments.

Lesson One – The Framework for Interpreting Our National Parks

Word challenges: framework, naturalistic (naturalism), worldview, assumption (assumes), secular, uniformitarianism, catastrophism (catastrophic), chronology

The National Park system – what a great idea! It has been the model for all nations since the inception of the first national park – Yellowstone National Park in 1872. The United States National Park system is the envy of the entire world, as people have been dedicated to preserving the wonders of nature for future generations to visit and study. And this was the intent when certain men petitioned Congress to protect the wonders of the Yellowstone region. It remains the primary goal of the National Park system today. I am so thankful that American people over 100 years ago had the foresight to commit themselves to this ideal.

What is a *framework*? A framework is an interpretation of evidence. Every human being uses a framework when thinking about life, science, history and philosophy. But did you know that *every scientist* also uses his framework to formulate ideas and opinions about physical evidence? And scientists are often unaware of just how their framework affects their conclusions. A good scientist will be aware of his framework and will be careful to keep that framework from influencing the results of research.

Sadly, the official interpretation or framework of our national parks has been taken over by a *naturalistic worldview*. That means that all the natural wonders enshrined in our national parks have been explained from a *secular* perspective so that the Biblical history of the earth is obscured. Many people are stumbled by this viewpoint, such that the book of Genesis is becoming a just-so story not rooted in reality. Most people don't give it a second thought. However, the implications for the Christian are enormous. Everything that has been taught about Jesus Christ is rooted in the book of Genesis. If Genesis is false, then the foundation for Jesus as our Messiah and Savior is suspect at best, and a lie at worst.

Early in the development of modern geology, men adopted a naturalistic framework for interpreting the landforms of our earth called *uniformitarianism.* This initially meant that the geological forces that are shaping our earth today have been going on almost indefinitely. A phrase used by Dr. Gary Parker states this view in a way that is easy to remember: "*Small and slow and long ago*". Although the term *uniformitarianism* has been slightly modified today to include localized geological catastrophes, like volcanic eruptions, the concept of an ancient earth, perhaps as old as 4.6 billion years has remained. This means that when a secular geologist looks at Yellowstone Park, for example, because he sees slow and gradual erosion in operation today, he *assumes* that this geological force has been going on for hundreds of millions of years. Those who formulated modern geology in the early 1800s made a huge historical, and consequently a huge geological, mistake. They rejected the global Flood of Genesis. If indeed a global flood had occurred as Genesis records, this concept would have totally changed the course of modern geology. The global Flood of Genesis would have produced the effects highlighted in another one of Dr. Parker's sayings, *"Big and fast, and in the recent past."* These two sayings are diametrically opposed to each other.

Every one of our national parks are interpreted through the secular, uniformitarian framework to tell the public a naturalistic story with a naturalistic beginning.

What a difference it makes when a person takes Genesis as recorded history and interprets our national parks through this framework. The Bible seems to come alive and the reader is usually struck with a fresh sense of the historicity and reality of the Biblical record.

In contrast to the uniformitarian framework involving *small and slow* geological processes over millions of years, is the Biblical view or the *catastrophic* framework of Earth history. The first 11 chapters of the Bible give a concise and historical framework through which the landforms of the Earth can be interpreted. It is best summed up in the diagram below.

A Basic Biblical Framework

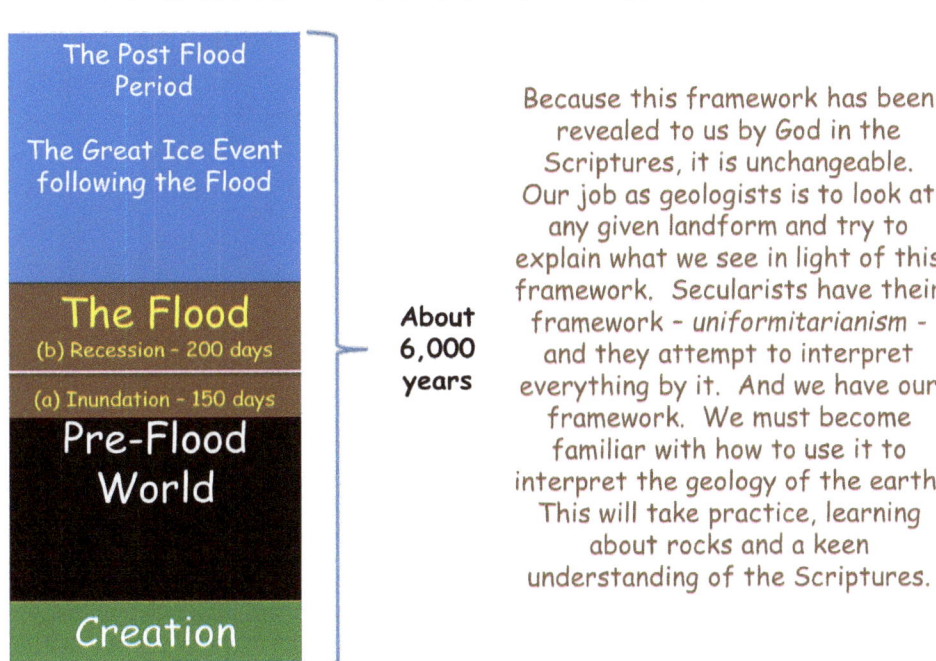

In this framework, we don't start with evidence, because a person will always be influenced in the interpretation of that evidence by his philosophical framework. Instead we start with a recorded history of the Earth. The Biblical, historical framework has been tested historically, geographically and archaeologically, time and time again and has stood the scrutiny of thousands of people through the years. This framework is what we will use to interpret the evidence. It helps us to fit what we observe into a more historically complete picture. Modern geology tells their own historical story in interpreting the geology of our national parks and then weaves evidence into that story. This will almost always be contrary to the Biblical story. And it is this secular approach that has caused so much confusion and consternation for those who believe the Biblical story.

The Bible also reveals a historical *chronology* of the Flood in Genesis chapters 7-8. Some may offer a little different analysis of the number of days involved in the Flood period, so the following chart is simply my attempt to promote a general geological structure of the period known as The Flood.

The Genesis Flood – Order of Events and the Geological Implications

(not to scale) Genesis 6:9-8:16, beginning about 4,500 years ago

Gen. 6:10-22, Noah receives the commission to build an ark (box) out of gopher wood with three decks and rooms sealed with tree resin (pitch); 450' long, 75' wide and 45' high; he was to gather pairs of all land-dwelling, air-breathing animals, birds and creeping things of the ground and put them in the ark because God would soon destroy the earth with a flood.

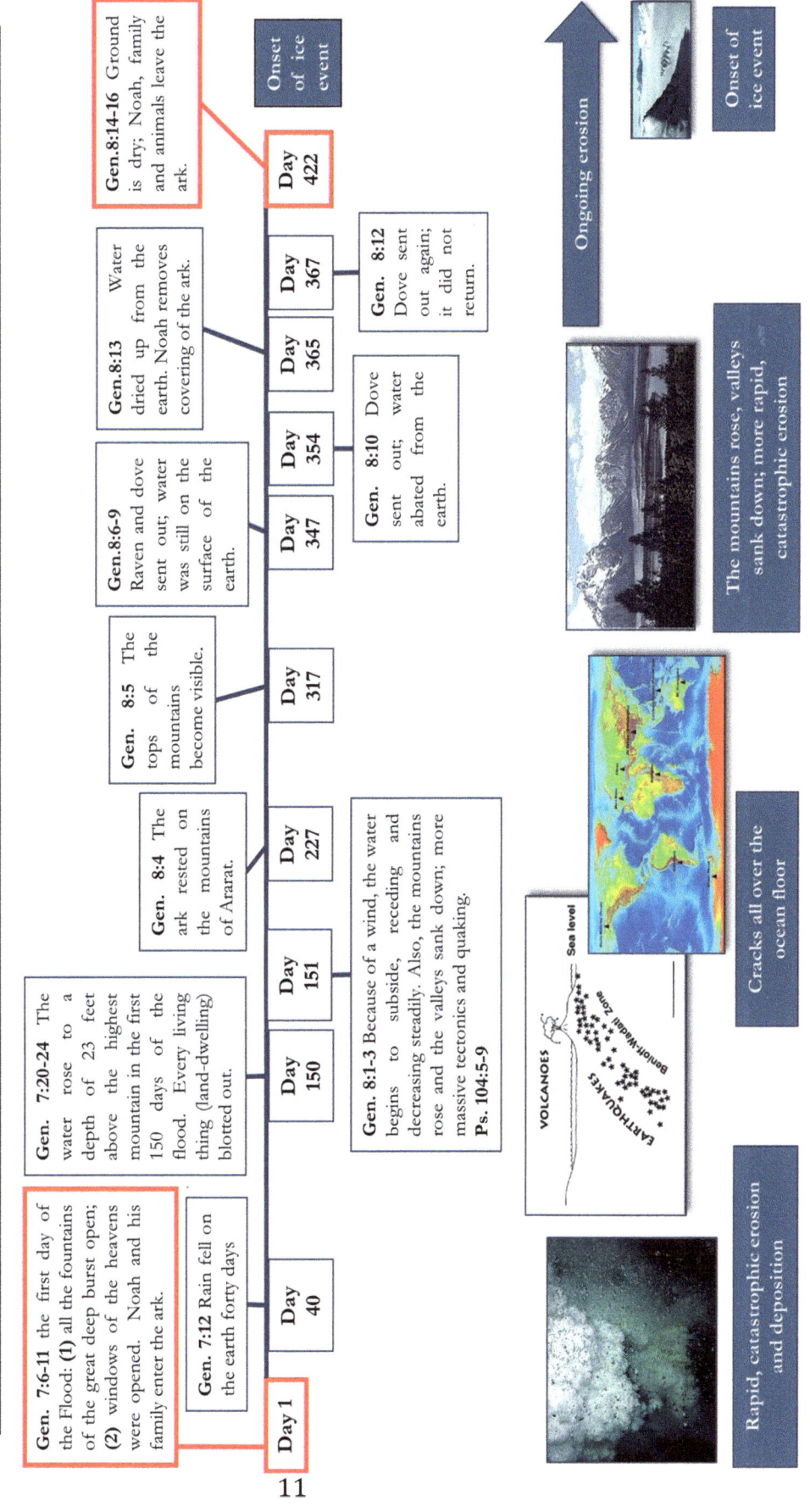

Day 1 — **Gen. 7:6-11** the first day of the Flood: (1) all the fountains of the great deep burst open; (2) windows of the heavens were opened. Noah and his family enter the ark.

Day 40 — **Gen. 7:12** Rain fell on the earth forty days

Day 150 — **Gen. 7:20-24** The water rose to a depth of 23 feet above the highest mountain in the first 150 days of the flood. Every living thing (land-dwelling) blotted out.

Day 151 — **Gen. 8:1-3** Because of a wind, the water begins to subside, receding and decreasing steadily. Also, the mountains rose and the valleys sank down; more massive tectonics and quaking. Ps. 104:5-9

Day 227 — **Gen. 8:4** The ark rested on the mountains of Ararat.

Day 317 — **Gen. 8:5** The tops of the mountains become visible.

Day 347 — **Gen. 8:6-9** Raven and dove sent out; water was still on the surface of the earth.

Day 354 — **Gen. 8:10** Dove sent out; water abated from the earth.

Day 365 — **Gen. 8:13** Water dried up from the earth. Noah removes covering of the ark.

Day 367 — **Gen. 8:12** Dove sent out again; it did not return.

Day 422 — **Gen. 8:14-16** Ground is dry; Noah, family and animals leave the ark. **Onset of ice event**

- Rapid, catastrophic erosion and deposition
- Cracks all over the ocean floor
- The mountains rose, valleys sank down; more rapid, catastrophic erosion
- Ongoing erosion
- Onset of ice event

We will use *A Basic Biblical Framework* discussed previously, and this chronology as we look at, and seek to interpret, several of our national parks.

Thought Questions
(most answers and references can be found in the text)

1. Define the word, *framework*.

2. Define the word, *uniformitarianism*.

3. Define the word, *catastrophism*.

4. Define the word, *secular*.

5. Dr. Gary Parker uses two sayings to summarize the two opposing frameworks used in interpreting the geology of the Earth. What are they?

6. Briefly list the four main parts to the Basic Biblical Framework.

Activity: Memorize the Basic Biblical Framework above and then draw it from memory. It is vitally important to know this framework if we are to understand the geology of our national parks.

Lesson Two – Our Restless Earth:
Yellowstone National Park

Word challenges: fumarole, geyser, neocatastrophism, caldera, fault, earthquake, magma, earthquake swarms, tectonics, rhyolite, tuff, geyserite, sinter, travertine

Yellowstone National Park

Yellowstone National Park and its surrounding areas is a veritable laboratory of examples of catastrophic geological events. These include several **caldera** eruptions in Wyoming and Idaho, massive mountain uplift of the Beartooth Mountains in Montana and Wyoming, the largest accumulation of petrified trees in ash in the world in the Gallatin Mountains, hundreds of **earthquakes** every year, unbelievable evidence of immense glaciers that once occupied the Yellowstone area, and the breathtaking volcanic remnants that form the Absaroka Mountains surrounding Yellowstone National Park on two sides. All these features defy a uniformitarian explanation. They are best explained by a horrendous global, catastrophic event recorded in the seventh and eighth chapters of the Book of Genesis.

Yellowstone: A Short History

On April 30, 1803, the nation of France sold 828,000 square miles of land west of the Mississippi River to the young United States of America in a treaty commonly known as the Louisiana Purchase. The United States paid $11,250,000 for this chunk of real estate! President Thomas Jefferson was eager to know about his purchase and what lay beyond the Mississippi River. It was truly a frontier and only a few trappers and traders had brought back reports of this amazing country that lay beyond the Mississippi River.

In 1804 President Jefferson enlisted Lewis and Clark to explore and map the Louisiana Purchase and to seek a route to the Pacific Coast. The Lewis and Clark Expedition became known as The Corps of Discovery. It took them two years to complete this journey and they brought back an amazing amount of geographical, biological and geological information. Although they had heard rumors of a fantastic area close by that had been reported to have spouting hot water and beautiful waterfalls, boiling hot springs and mud volcanoes, they did not explore to find out if it was true.

It was on the return trip of the Lewis and Clark Expedition that one of the hired guides decided for some reason to break off from the Corps of Discovery and follow the Yellowstone River into what is now Yellowstone National Park. His name was John Colter. He brought back fantastic tales of volcanic scenes that captured the imagination of many people. Others thought he was crazy and his descriptions of this amazing place earned it the name, Colter's Hell. He also told of seeing huge numbers of buffalo, moose, elk, and beaver. This caught the attention of the *mountain men*. And these men struck out for the new Yellowstone country to earn their fortunes in the fur trade.

The abundant buffalo that Colter saw have only started to return to their original numbers in the last 40 years.

The first major US government-backed scientific expedition to the Yellowstone area was undertaken in 1871 and is known as the Hayden Expedition, after its leader Ferdinand Hayden, geologist. Darwin's *Origin of the Species* had been in print since 1859 and modern geology was no longer in its infant stages. The Biblical framework had long been abandoned and like most geologists of the times, Hayden had been instructed in the uniformitarian framework as the only scientific way to interpret the Earth's geology. Hayden's reports and influence led Congress to declare the Yellowstone region a protected area and it became the nation's first national park in 1872. In subsequent years following the Hayden Expedition, geologists began to study and map the various features of the new Yellowstone National Park that are now visited by over three million tourists a year.

Yellowstone National Park represents one of the finest efforts by man to preserve a wonderland of geological and biological phenomena. Located in the three states of Wyoming, Montana and Idaho, it comprises 3,468 square miles, a little over three times the size of the State of Rhode Island. A recent survey showed that over 1,200 *geysers* have erupted in Yellowstone. To date there are over 10,000 thermal features, 300 waterfalls and a variety of plant and animal life that have been catalogued in this amazing place.

Up until the late 1960s, secular geologists saw Yellowstone as simply a wonderful collection of isolated hot springs, *fumaroles*, mud pots and geysers. With the advent of new ideas like *neocatastrophism* and the development of modern aerial photography, Yellowstone began to be seen as a *caldera* – a massive volcano that had exploded at some point in Earth's history, leaving its remnant of thermal features, *faults*, and *earthquakes* all seeming to be connected to a *magma* chamber several miles below the Earth's surface.

- Hot spring – hot water from deep underground travels up through cracks in the *rhyolite* rock without constriction
- Mud pots – acid-rich hot springs where the acid has turned the *geyserite* into gray mud
- Fumarole – steam vent only, very little water released, but it makes a hissing sound
- Geyser – the eruption of hot water through constricted rhyolite vents

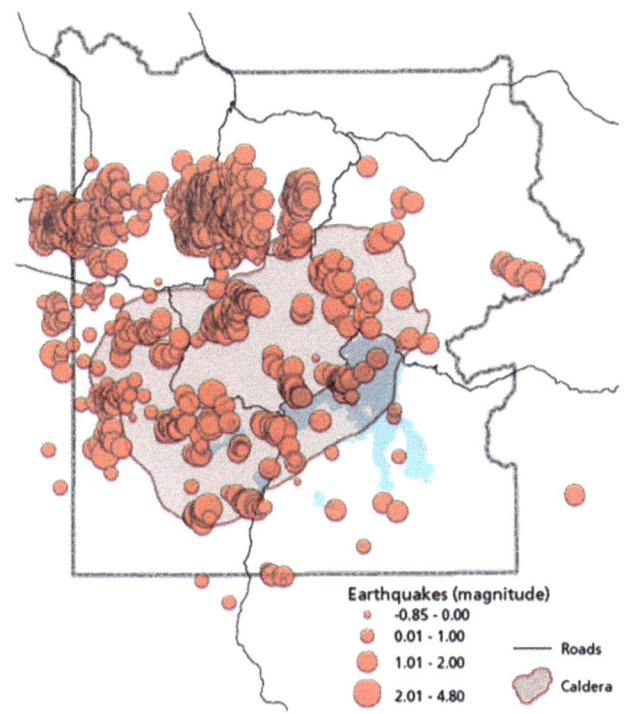
Map charting the recorded earthquakes and their magnitude in Yellowstone

Earthquakes of Yellowstone

Yellowstone commonly experiences *earthquakes*, sometimes daily! Approximately 2,000-3,000 earthquakes occur in Yellowstone each year. *Earthquake swarms*, a fairly new term in the Yellowstone literature, represent a series of earthquakes over a short period of time in a localized area. The largest swarm occurred in 1985, with more than 3,000 earthquakes recorded during three months on the northwest side of the Park. Hundreds of quakes were recorded during swarms in 2009 (near Lake Village) and in 2010 (between Old Faithful area and West Yellowstone). Scientists interpret these swarms as due to shifting and changing pressures in the Earth's crust that are caused by migration of hydrothermal fluids, a natural occurrence of volcanoes. The real question to this phenomenon, however, is why the Earth experiences earthquakes in the first place.

Earthquakes occur because of fractures in the Earth. These fractures in the Earth's crust are called *faults*. These faults shift and indicate that the Earth is weak geologically and therefore restless. At times, there are secondary forces that cause these fractures to shift such as shifting hot liquids in the Earth's crust. Secular geologists view faults as a normal part of Earth behavior which has been going on for 4.6 billion years – a natural part of Earth's geological evolution. That is a naturalistic worldview and embraced by most people in the world today.

The primary cause of earthquakes, however, within a Biblical framework, was the historical Flood of Genesis chapters 7-8. In Genesis chapter 7:11 we read that on the first day of the Flood, "...*the fountains of the great deep burst open.*" The straightforward implication of this global geological event was the initiation of cracks (faults) in the Earth's crust which would have produced magma eruptions, release of hot water and steam, radioactivity, volcanic eruptions, and the permanent shifting in the Earth's crust we now experience as earthquakes. Some areas of Earth are more fragile than others. But all the Earth was affected and continues to be affected by this one, global geological event – The Genesis Flood.

Black Dragon's Caldron – a modern example of the on-going effects of a restless Earth

But it isn't just the earthquake swarms that change the geological landscape of Yellowstone. In 1959 the largest earthquake in the Rocky Mountains ever recorded caused part of the Madison Mountains to give way. This area is just outside the Yellowstone National Park boundaries, northwest of the park. A 7.3 magnitude earthquake caused an 80-million-ton landslide to form a new lake, dropping Hebgen Lake Dam by 11 vertical feet. This earthquake caused changes in the park. Some of the thermal features stopped erupting, some features erupted with more intensity, and others that had been silent for years suddenly came to life.

This is the site of the landslide that damned the Madison River and created present-day Earthquake Lake. This picture is of the mountain face, looking across the canyon. The center of the picture is where a large mass of dolomite rock detached. Eighty million tons of debris rolled down the mountain, blocking the Madison River, and came to rest on the other side of the canyon, with about 100 vertical feet of rock debris.

The 1959 earthquake that shook the Hebgen lake area and sent this house into the lake was in the middle of the night. Those vacationing in the area had to scramble from their homes and tents to escape. This house broke off from its foundations, and floated to its present location.

Plate Tectonics and Yellowstone

Tectonics is from a Greek word, meaning, *pertaining to building*. What we observe at present are earthquakes, faults, volcanic eruptions, and tsunamis all of which produce shifting in the Earth's crust. We don't witness any constructive building. Within a Biblical framework all this geological activity can be related back to the one global event, the Genesis Flood. Secular geologists view tectonics as a continuing geological process that began 4.6 billion years ago and tectonics has naturalistically built our Earth from one land mass to several land masses. And tectonics has done this at least twice in Earth history, according to secular geologists. This idea, developed in the 1970s, is now termed *plate tectonics*. It has not been shown to be true to the extent that secular geologists ascribe shifting continents to it. But because of secular geologist's rejection of the Biblical Flood, plate tectonics is the only mechanism in their understanding that explains how we got the individual continents of today. But there is another possible view! The Genesis Flood would certainly offer a different interpretation for the origin of the continents. If we start with the one land mass as taught in Genesis chapter one at creation and then apply the tectonic upheaval caused by the Genesis Flood, then we don't have the earth *building* itself together, but *being ruined*. The earth of today is not the same Earth that once was. The Flood has produced our oceans and ocean basins and flooded huge portions of the one main landmass that was formed in Genesis chapter one, and washed out large chunks of the original landscape, leaving remnants of the original land mass of Genesis chapter one in the form of *continents*.

Our Restless Earth: Calderas

No caldera eruption has been observed in modern times on the scale of Yellowstone. Several caldera impressions have been discovered related to the Yellowstone area. The most conspicuous is the caldera occupying a good portion of Yellowstone National Park. Its crater has been measured at 45 miles long and 35 miles wide! This certainly fits the definition of *catastrophic*!

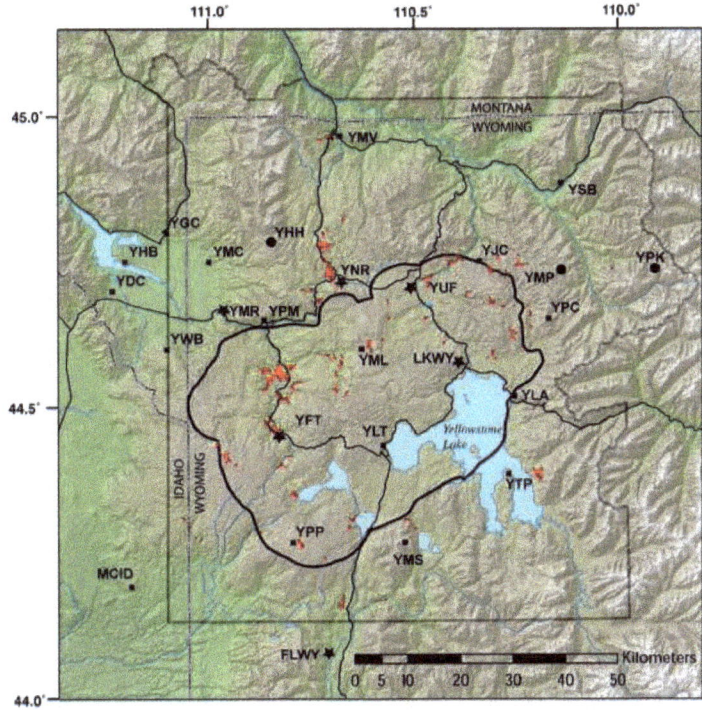

In the above map, the outline of the Yellowstone Caldera is in black. This map shows the seismic network of Yellowstone National Park. The black dots and stars are earthquake monitoring stations. The thick black line is the boundary of the Yellowstone Caldera. Gray outline is the park boundary. Red regions are thermal areas.

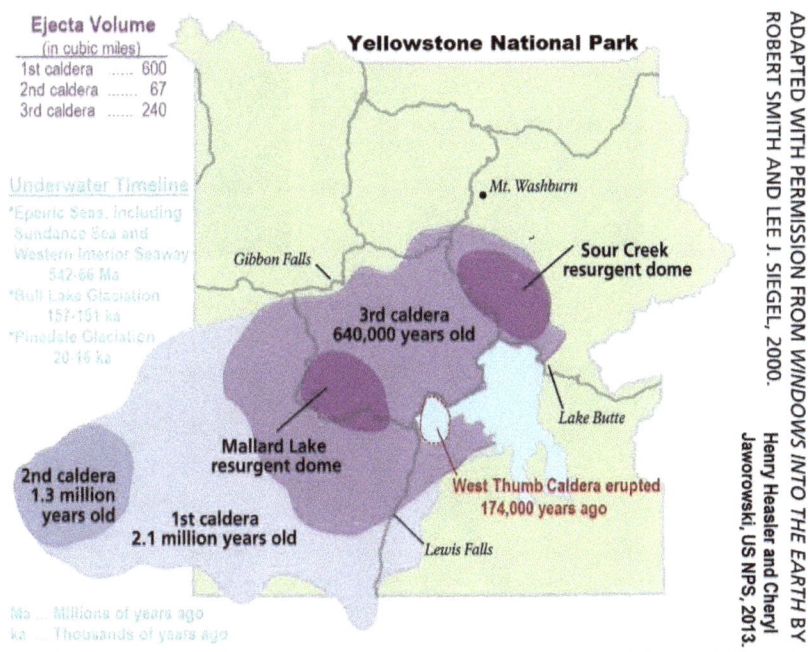

Although I am not in agreement with the secular ages of Yellowstone, evidence of six caldera explosions have been identified primarily through the chemical variations in the tuff that was erupted.

Volcanic Rocks of Yellowstone

Like other volcanoes Yellowstone displays lava flows and various other types of lava eruptions. What is unusual for volcanoes, however, is the mind-boggling amount of *rhyolite* lava, ash and rhyolite *tuff* spread out over Yellowstone – in some cases 100s of feet thick. Most volcanoes do not even come close to erupting this kind and amount of volcanic material. Rhyolite is high in quartz making it a product of a very explosive eruption. The amount of rhyolite indicates that Yellowstone was a violent geological event in Earth's history. (You can learn a bit more about this in Lesson Eight.)

Basalt columnar jointing at Sheepeater Cliff, one of the few basalt flows in Yellowstone

Rhyolite lavas and tuff – the most abundant volcanic rocks of Yellowstone National Park; they contain a great amount of quartz and are associated with explosive eruptions.

Geyserite or siliceous *sinter* - another common rock of Yellowstone; it is formed when hot acidic waters travel up through rhyolite and then precipitates out on to the ground and leaves a deposit of silica called sinter, typical with geysers and hot springs.

Riverside Geyser, one of 10,000 thermal features in Yellowstone.
Notice the geyserite or sinter mound from which the geyser is erupting.

Travertine, a type of limestone, from Mammoth Hot Springs, Yellowstone, is formed when hot acidic water travels through limestone and dissolves it. Upon reaching the surface of the ground, the hot water precipitates its load as a type of limestone called travertine.

Mammoth Hot Springs produces travertine because of the limestone foundation underneath Yellowstone in this area of the park.

Parting Shots

A visit to Yellowstone National Park is a life-changing trip. The geological evidence here cries out for a different interpretation of Earth history.

Over 10,000 feet of vertical uplift has taken place in the Beartooth Mountains. These mountains are made of the basement rocks, granite and gneiss. The tops of the mountains have been planed flat by some kind of rapidly moving force and then rounded from glaciers.

The Gallatin Range of mountains just to the northwest of Yellowstone contain the largest concentration of petrified logs in the world.

Glacial cirques are evidence of the occupation of once mighty glaciers in the Madison Range of mountains just northwest of Yellowstone.

Catastrophic glacial melting and flooding towards the end of the great ice event would have left behind these water tumbled, rounded cobbles in an unsorted sedimentary mix just below the Madison mountains.

The Absaroka Range of Mountains: over 9,000 cubic miles of volcanic material testifies to the catastrophic origins of Yellowstone.

The geology of Yellowstone National Park testifies to our restless Earth. The catastrophic nature of the Yellowstone evidence defies a naturalistic, uniformitarian explanation. In my opinion the better interpretive framework is the Basic Biblical Framework found in Lesson One.

Thought Questions

1. How are earthquakes and faults related? From a Biblical framework, explain the origin of faults.

2. What is the difference between modern volcanic eruptions and those of Yellowstone?

3. Name the four kinds of thermal features in Yellowstone and briefly describe the differences.

4. Describe some of the geological features in and around Yellowstone that are better explained by a catastrophic origin.

5. Explain the difference between present faulting combined with earthquakes, and plate tectonics as envisioned by secular geologists.

Activity: Obtain a copy of my book, The Geology of Yellowstone to get a deeper understanding of the geology of the Yellowstone Ecosystem.

Lesson Three – Fossils, Preservation and Petrification:
Petrified Forest National Park, Hagerman Fossil Beds National Monument, John Day Fossil Beds National Monument, Fossil Butte National Monument, and Badlands National Park

Word challenges: petrified (petrification), fossil, conglomerate, silica, quartz, lungfish, disarticulated, planation

Petrified Forest National Park

Located in northeastern Arizona, Petrified Forest National Park comprises 230 square miles and is known for its large deposits of *petrified* wood. The word petrify means, to turn to stone. Today, the word fossilize and petrify are used interchangeably. A *fossil* is simply the remains of past life, turned to stone. Petrified Forest National Park became a national park in 1962 after being upgraded from its original designation in 1906 as a National Monument.

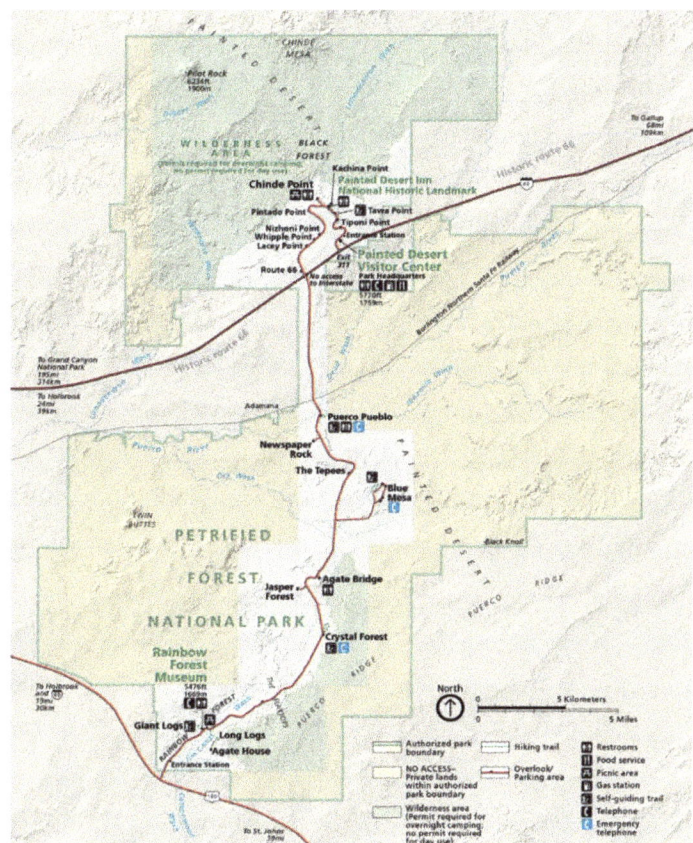

Map of Petrified Forest National Park

Petrified Forest National Park is part of a much larger geological feature known as the Colorado Plateau, a vast area of uniform rock type, measuring over 130,000 square miles within western Colorado, northwestern New Mexico, southern and eastern Utah, and northern Arizona. Much of the Colorado Plateau is characterized by flat, planed surfaces, known as *planation* surfaces. The mystery concerning planation surfaces is determining what geological force would have created broad, flat surfaces in the midst of deep cut gorges. The Genesis Flood would have involved the formation of flat, planed surfaces through sheet erosion in the initial receding floodwaters. The gorges and the canyons would have been cut rapidly by the channelization of the final receding floodwaters.

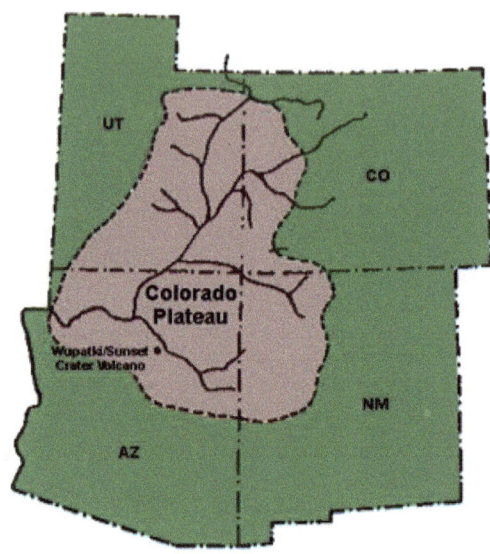

The uniformitarian framework presents Petrified Forest National Park as preserved remains of the Triassic Period of geological time, about 250 to 200 million years ago, an interval covering about 50 million years of Earth history. The Triassic is also called the Dawn of the Dinosaurs. But how do secular geologists arrive at these ages? No *fossils* are radiometrically dated. These dates are arrived at using the evolutionary idea that life suddenly appeared about 550 million years ago and quickly evolved and diversified. However, the Biblical framework for the geology of the area will readily show things in a radically different perspective.

A Rocky Forest
The kinds of rocks covering the Petrified Forest National Park are primarily volcanic ash, clays, sandstone, and ***conglomerate***. These all give evidence that some catastrophe happened here in the not so distant past. The conglomerate consists of water tumbled, rounded stones. The wood in the petrified logs has all been infiltrated by ***quartz***, most likely derived from the ***silica*** of the volcanic ash in what was once a tremendous amount of water. Secular scientists are divided over the length of time involved in producing this fossilization. Many believe that it took millions of years to accomplish this. But there is a problem with this - it fails to account for decay that would have set in over that time. There is a very good explanation that does not require millions of years, but the right

environment for this change. The process would have included rapid burial and infiltration of silica-rich solutions, that would have quickly turned the trees to stone. This could have easily taken place during the time of the Genesis Flood. Slowly petrifying logs over millions of years will not explain the beautifully preserved varieties of wood found in this wonderland of fossils.

One of the petrified logs of Petrified Forest covered with conglomerate, indicating a water-tumbled burial

Close-up of the conglomerate that covers the log

One of the many sandstone formations of Petrified Forest showing the preservation of water currents and conglomerate

Rounded stones cover the floor of Petrified Forest. These indicate a watery deposition and fast-moving currents.

One of the thousands of petrified logs resting on top of the conglomerate floor of Petrified Forest

Some of the eroded features of volcanic ash deposits that characterize the Petrified Forest National Park and the Painted Desert to the east of the park

When is a Forest Not a Forest?

Petrified Forest National Park may not be a fossil forest at all, but are, in fact, petrified remains of trees and other fossils that were part of a catastrophic watery transportation, deposition and burial of once living things and sediments. Because of the watery transport and depositional characteristics of Petrified Forest National Park, the original location of the fossils remains a mystery. Who knows where they originally came from?

Strange Creatures in the Forest
Petrified wood is not the only fossil found in this national park. A variety of plant fossils, including ferns, cycads, ginkgoes, and many other plants have been discovered. Many insects, reptiles, amphibians, and fresh water fish have been catalogued, including freshwater sharks and a myriad of fishes like coelacanths and other *lungfish*. Even mammals are among the fossils. Invertebrates, including freshwater snails and clams have also been discovered. These fossils all indicate a tumbled, mixed mess of remains that seem to have been washed into place in raging floodwaters. These remains were most likely deposited in floodwaters that were as a result of the end stages of the Genesis Flood, as sedimentary earmarks comprise the predominate features of the surface area.

Fossil of a Phytosaur, a type of crocodile found in Petrified Forest National Park

Reptiles Found in the Area

The dinosaur, Coelophysis, is also found in Petrified Forest National Park.

One of the most apparent observations concerning Petrified Forest National Park is the *disarticulated* state of the fossil logs. (Disarticulation is the state of being broken up and torn apart.) They are for the most part broken up and scattered over the surface of the terrain. These, along with the abundance of planed surfaces of the many plateaus that are part of the park, all indicate the past action of rapidly moving water.

Notice the disarticulated state of most of the petrified trees of the park. Notice, too, the planed surfaces in the background. These indicate a large amount of rapid water erosion over a short period of time.

Thought Questions

1. Why is it not appropriate to call this particular park a petrified forest?

2. What is thought to take place in petrification?

3. What is the mineral that replaces the original cells of the wood? Where did it come from?

4. What varieties of fossil animals and plants are found in Petrified Forest National Park?

5. How old do secular geologists say Petrified Forest is? What geological period do they call this time?

6. Briefly describe the formation of Petrified Forest National Park using the Basic Biblical Framework.

Activity: List some of the fossilized plants and creatures found in Petrified National Forest. What do uniformitarian geologists say about them? How is that similar or different from a Biblical framework?

Hagerman Fossil Beds National Monument

Word challenges: clastic, tephra, silicic, vertebrates, attrition

The Hagerman Fossil Beds are located southeast of Boise, Idaho, about 108 miles. They contain the richest ice age fossil beds in the world. The Hagerman fossil beds were declared a National Monument in 1975.

The Hagerman Fossil Beds National Monument is located along the Snake River.

The rocks that make up the Hagerman fossil beds are sedimentary, primarily sandstone, which is a *clastic,* water-formed, sedimentary rock. Geologists call the clastic sedimentary rocks *clastic sedimentary packages*. The word clastic means *broken* and refers to the tiny bits and pieces of rock and mineral, cemented together to form sedimentary rocks. A package refers to what is called a **suite** of various rock formations. These clastic sedimentary packages are interbedded with basaltic flows, pyroclastic ***tephra*** and ***silicic*** volcanic ashes, have a cumulative thickness of 5,000 feet, and comprise the seven formations of what is called the Idaho Group. What could explain this incredible sedimentary suite of rocks, but a catastrophic event of some kind.

Glacial Lake Bonneville
The Hagerman Fossil Beds fit geographically in the same vicinity of the Glacial Lake Bonneville Flood. Geologists believe that Glacial Lake Bonneville was created after the most recent ice age, in the Pleistocene Epoch. As a creationist, I view this, instead, as taking place toward the end of the ice event following the Flood. During this time, Glacial Lake Bonneville, stretching all the way from southern Idaho to the modern Salt Lake City area, reached its highest level of water. It is believed that a large amount of earth that was holding back the northern most water level broke open and created the Lake Bonneville Flood. When the dam collapsed, it is estimated that it released a 410-foot flood crest down the Portneuf River Valley, in what is now Idaho, also spilling into the neighboring Bear River Valley. When it reached the Snake River, it eroded away a lava dam that had been at

the site of the present-day American Falls, releasing a 40-mile long lake, American Falls Lake, that had formed behind the natural dam.

Red Rock Pass, supposed sight of the earth dam breach

Path of the Lake Bonneville Flood that moved thousands of heavy basalt boulders, tumbling them in the process

One of the tumbled basalt boulders from the Lake Bonneville Flood

Hagerman, Idaho; notice the car key for scale.

Map showing the extent of some of the post ice age floods that were the result of ice and rock dam breaches

Fossils found at the Hagerman Fossil Beds

The Hagerman Horse: the largest concentration of horse fossils is found at the Hagerman Fossil Beds, primarily consisting of 200 disarticulated skulls

The famous fossils found in the Hagerman Fossil Beds are a product of the catastrophic Lake Bonneville Flood. The geological explanations for the formation of the Hagerman fossil beds have been extremely varied ranging from *attrition* of the animals found there to catastrophic flooding, from one flood event to many flood events. One of the best examples of this from the area is the Lake Bonneville Flood, believed to be the second largest ice age flood in known geologic history. Certainly, the Lake Bonneville Flood could have accounted for the vast array of fossils that are found here.

Besides the fossil horse, other large *vertebrates* collected from the ***quarry*** include an antelope, a camel and a peccary. Small fossil mammal fossils found here include hare, weasel, gopher, vole, and shrew. In addition, there are fossils of woodland birds, waterfowl, snakes, turtles, lizards, frogs, toads, salamanders, a variety of fish, bivalves, and gastropods. It is a mix of habitats and indicates some kind of catastrophic, watery mix, perhaps from one of the many post glacial events that followed the Flood.

Thought Questions

1. What might the Hagerman Fossil Beds have to do with the Lake Bonneville Flood?

2. What is the significance of the disarticulated skulls found at Hagerman Beds?

John Day Fossil Beds National Monument

Word challenge: breccia

John Day Fossil Beds National Monument is known for its well-preserved layers of fossil plants and mammals. It consists of 14,000 acres in northcentral Oregon and was established as a national monument in 1975. The park headquarters and main visitor center are 122 miles northeast of Bend and 240 miles southeast of Portland by Oregon State Highway 218.

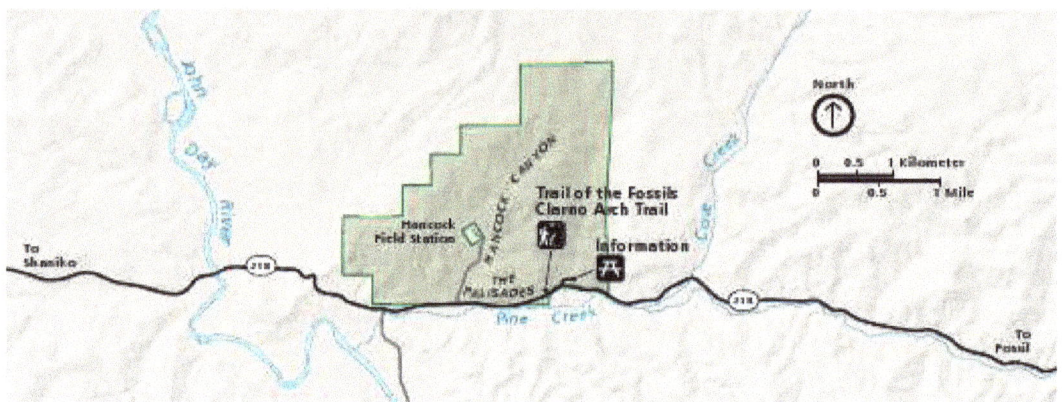

John Day Fossil Beds National Monument located along Oregon Highway 218.

Within the John Day Monument is a formation called the *Clarno Nut Beds*, a geological formation containing over 175 species of nuts and fruits preserved in volcanic lahars. These fossils are of tropical and subtropical nuts, fruits, roots, branches, and seeds. The Clarno Formation also contains bones, palm leaves longer than 24 inches, avocado trees, and other subtropical plants. The fossils indicate a climate that was warmer and wetter than it is in the 21st Century. Secular geologists interpret this as a by-gone environment that used to characterize this area of Oregon. Large mammals preserved in the sedimentary rocks include brontotheres and amynodonts, scavengers like the hyaenodonts, as well as Patriofelis, and other predators. Eroded remnants of the Clarno stratovolcanoes, once the size of Mount Hood, are still visible near the monument - for example Black Butte, White Butte, and other buttes near Mitchell, Oregon.

The environment that supposedly existed in this part of Oregon at one time in the past

Black Butte, Oregon

A Brontothere

As with most localities that display fossil beds, the John Day fossil beds have a watery, catastrophic origin consisting of volcanic eruptions which deposited lavas, accompanied by debris flows (lahars) atop the older rocks, in the western part of the province. Containing fragments of shale, siltstone, conglomerates, and **breccias** (which are all sedimentary rocks), the debris flows entombed plants and animals caught in their paths. It is most likely that these creatures were brought into the area by these lahars. Secular geologists date these formations at around 50 million years old based on the radiometric dates of the volcanic rocks of the region. Using our Biblical historical framework, however, there isn't that kind of time available. The Biblical framework might place these deposits toward the end of the Flood when, according to Psalm 104:5-9, "The mountains rose; the valleys sank down...." There would have been on-going volcanism as a result of this geologic activity, as well as ash clouds and residual flooding of local areas. The mix of tropical and subtropical fossils discovered would also indicate a stirring together of animal and plant remains in a slurry of mud and volcanic ash, which contains a high amount of silica. The silica would have greatly aided rapid and complete fossilization before decomposition.

Other fossils found in the John Day Fossil Beds include a wide variety of plants and more than 100 species of mammals, including dogs, cats, oreodonts, saber-toothed tigers, horses, camels, rodents, turtles, opossums, large pigs, rhinoceroses, bears, pronghorn, deer, weasels, racoons, and sloths. These are scattered across a wide area in the park. More than 60 plant species are fossilized in these strata, such as *hydrangea*, peas, hawthorn, and mulberry, as well as pines and many deciduous trees. One of the notable plant fossils is the *Metasequoia* (dawn redwood), a genus thought to have gone extinct worldwide until it was discovered alive in China in the early 20th century. It is a genuine *living fossil*.

Strata typical of the John Day Fossil Beds National Monument

Kinds of fossils preserved in the John Day Fossil Beds

Brontotheres - rhino-like animals

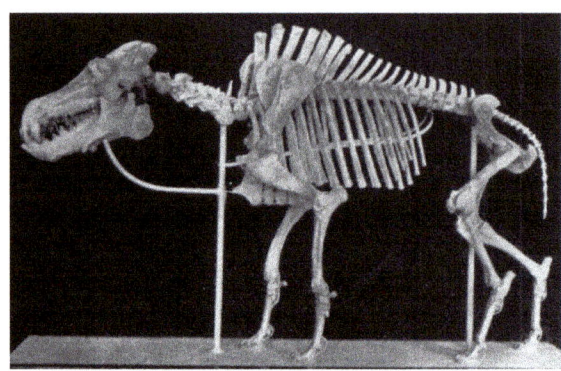
Entelodonts – giant pigs

Hippo-like animals and hyena-like animals

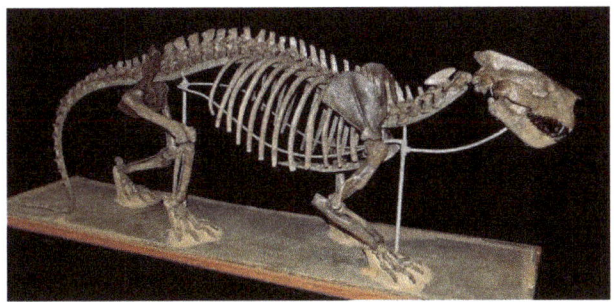

Tiger-like animals

Oreodont: most likely a sheep-like animal

Secular geologists imagine an environment that once lived in the present John Day area of what is now Oregon, based on fossils they discover. But this is imagination, not science. The exact history of these animals cannot be tested by science. They must be interpreted by a worldview. Both involve history, science and philosophy. But insisting that these animals lived here is not scientific fact. It is an idea, at best. They could have just as easily been washed in during the Flood or post-Flood.

Imaginative environments supposedly based on fossils found at John Day; but this presupposes that all these creatures lived here, and that their remains were not washed in.

Thought Questions

1. What could be some reasons not to view the John Day Fossil Beds as preserved past environments?

2. What is the relationship between volcanism and the preservation of the fossils at the John Day Fossil Beds?

Fossil Butte National Monument

Word challenge: varve

The National Park Service description of Fossil Butte states, *"Some of the world's best preserved fossils are found in the flat-topped ridges of southwestern Wyoming's cold sagebrush desert. Fossilized fishes, insects, plants, reptiles, birds, and mammals are exceptional for their abundance, variety, and detail of preservation. Most remarkable is the story they tell of ancient life in a subtropical landscape."*

Fossil Butte National Monument is best known for its millions of fossil fish buried in fine-grained limestone, showing exquisite detail.

Fossil Butte National Monument is located 15 miles west of Kemmerer, Wyoming, in southwestern Wyoming. It was established as a national monument in 1972. Fossils preserved at Fossil Butte include fish, alligators, bats, turtles, small horses, insects, and many other species of plants and animals. It is an aquatic mix of several habitats. Secular geologists interpret these fossils as belonging to three freshwater lakes with sediments that accumulated over a two-million-year period. The lakes supposedly covered parts of southwest Wyoming, northeast Utah and northwestern Colorado. The biggest of these lakes, Fossil Lake, was thought to be 40 miles by 50 miles. These are not small lakes!

Fossil Butte National Monument

Varves

The interesting mix of fossils could also be interpreted within the Biblical framework as having been washed in by retreating floodwaters or post-Flood localized flooding, remnants of which were concentrated in this geographical area. The fossils are found in fine-grained limestone with very fine and thin laminar layers that must have been laid down quite rapidly, as many of the fossils are preserved *through* the layers. The layers are said by secularists to represent *varves*, the annual light and dark seasonal deposits that are observed in present day lakes. But the fact that many of the fossils are found preserved throughout multiple layers betrays this idea. They must have been laid down rapidly with very little time in between deposits. If these are varves, then how would we explain preserved remains running through layers that supposedly represent years of sediment accumulation?

The fossil preservation is exquisite, preserving many fine details, indicating relatively little time passed between burial and preservation.

Insect fossils from the Fossil Butte area

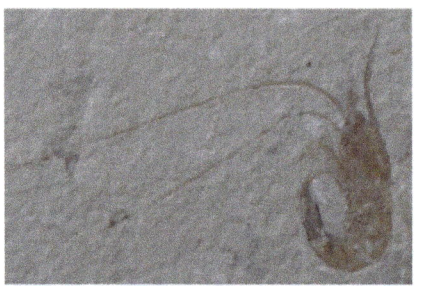 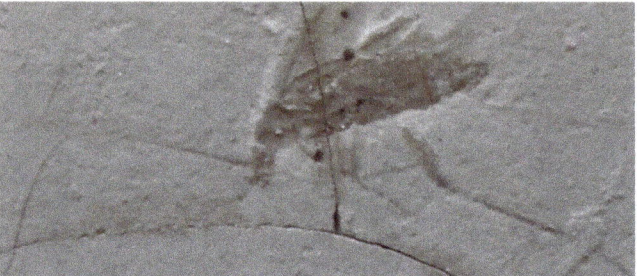

Shrimp fossil and insect fossil from the Fossil Butte area

Fossil fish

Fossil leaf

Fossil palm leaves, water plants and fossil fish from the Fossil Butte area

Fossil fish from the Fossil Butte area

Thought Questions

1. What are the possible interpretations for varves?

2. What would be the significance of the preserved fish throughout multiple layers of sediment?

Badlands National Park

Word challenges: clay, scute, micro-vertebrates, siltstone, erosion, deposition, ecological zone

Badlands National Park is located in southwestern South Dakota and comprises 379 square miles of intensely eroded *clay*, sandstone and volcanic ash. The movie, *Starship Troopers* (1997), was filmed there because of its "other-world" appearance. The clay is formed from altered volcanic ash involving heat and water.

The Badlands, as the park has traditionally been called, are part of a much larger area known as the White River Badlands that cover areas in South Dakota, Wyoming, eastern Colorado and Nebraska. Badlands National Park was declared to be Badlands National Monument in 1929 and changed status to a national park in 1978. The area contains an abundance of fossils. This has been known to Native Americans throughout their history. The Lakota Sioux knew about large fossilized bones, fossilized seashells and fossil turtle shells. They realized that the area had once been under water, and that the bones belonged to animals, which no longer existed.

The typical spires and pinnacles of the Badlands National Park

The White River Badlands contains a huge assortment of fossils from extinct land mammals and land turtles to fossils from what is supposed to be the Cretaceous Period, long before large mammals were supposed to have arrived onto the scene. Some of these include sea creatures like ammonites, nautiloids, fish, marine reptiles and turtles. The list also includes bones from Triceratops and teeth from Edmontosaurus, an ornithopod dinosaur, turtle shell fragments, alligator and crocodile teeth and *scutes* (armor), plant fossils, Tyrannosaurus rex teeth, pieces of Dakotaraptor, Thescelosaurus, Pachycephalosaurus or *bone-heads*, Ankylosaurus, Troodontids, and unusual *micro-*

vertebrates. All these fossils indicate a massive bone bed of fossil finds buried in some kind of a watery, volcanic catastrophe.

The rocks of the Badlands, including *siltstone*, sandstone, ash, altered clay, and conglomerate, indicates a significant time of volcanism and rapid water and wind *erosion*. This area "bleeds" into the Hell Creek Formation of eastern and southeastern Montana where dinosaurs from the Cretaceous Period (the Age of Dinosaurs, according to secularists) have been abundantly excavated.

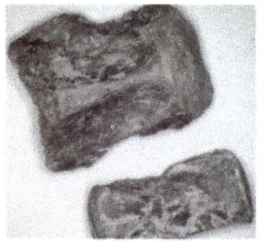 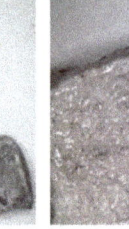 

Small mammal vertebrae   Turtle Shell   Ammonite

Dinosaur bones   Mammal coprolites

Large dinosaur vertebra

Teeth from a camel, and dinosaur fossils of the White River Badlands

Geologists date Badlands National Park as belonging to the period after the Cretaceous, that is, younger than 65 million years. Because these features are laterally aligned with other formations in the area, and are not isolated formations, they must be connected to a much broader area. The mix of bones and other fossils and the abundance of ash tells a story that differs from the uniformitarian view. The evidence better fits a catastrophic flood model which incorporates widespread volcanism; massive, rapid erosion; and transportation and *deposition* of sediments from a variety of environments. The badlands of Montana, North Dakota, South Dakota, Wyoming and Nebraska are all of similar composition. That fact turns these areas into a much broader, catastrophic event. A uniformitarian framework would interpret these things as representing localized flooding during the late Mesozoic and early Cenozoic Eras. What we observe there, however, really cannot be explained by localized flooding. It is too large an area. All the features are similar in rocks and terrain and all describe one tremendously catastrophic event.

Badlands of eastern Wyoming, southwestern South Dakota and eastern Montana

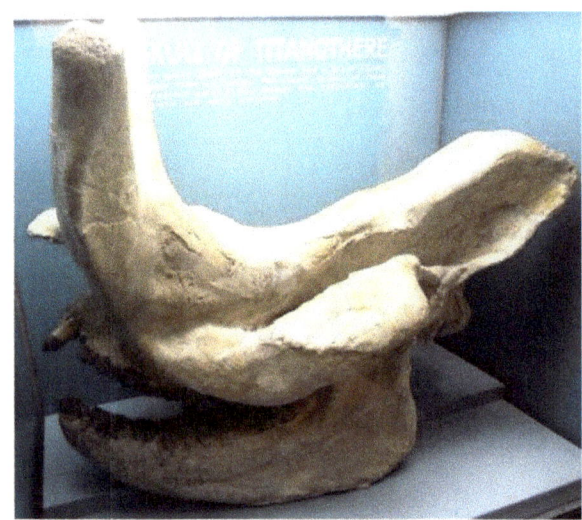
Fossil skull of a Titanothere, an extinct mammal of the White River Badlands

Mammals and Dinosaurs as Neighbors

The Titanotheres were huge mammals and they supposedly appeared on the scene immediately following the great dinosaur extinction of the Cretaceous Period. But they represent huge, almost unimaginable evolutionary leaps between dinosaurs and mammals. It seems more reasonable to think that these creatures lived at the same time as the dinosaurs, but in different *ecological zones*. This would explain why we generally do not find their bones mixed with dinosaurs. There are, however, exceptions to this. There were huge mammals found in the same dinosaur rock layers during Roy Chapman Andrews excavations in the Gobi Desert in China in the 1920s. Not much was ever said about this except for a few obscure articles of the time. But since then, more and more evidence has surfaced that places large mammals in the same rock layers as dinosaur.

Thought Questions

1. Describe the characteristics of the areas that are known as *badlands*.

2. What does clay have to do with the Badlands?

3. Describe both the Mesozoic Era and the Cenozoic Era.

4. List the animals that have been found in the sedimentary mix in the Badlands.

5. Why might we not generally find large fossil mammals buried with dinosaurs?

Activity: (Note: This activity is likely to be smelly!) Gather a variety of soils and sands. Put each of these in separate containers. Now, thoroughly soak the soils and sands and bury a piece of fresh uncooked meat in each container, being careful to act quickly and bury each piece thoroughly. Compress the sediments with a heavy object. Leave the heavy object in place. Put the container in a place that animals won't get to it. If you want, you can cover the container with a perforated lid. Wait a few weeks until the soils and sands are thoroughly dry. Next, wearing rubber gloves, dig up the pieces of meat and make your observations. What has happened to the meat samples?

Lesson Four – Stratovolcanoes, Andesite and Radiometric Dating
Mt. Rainier and Mt. Lassen National Parks

Word challenges: composite volcano, stratovolcano, pyroclastic, andesite, quartz, lahar, radiometric dates, assumption, isotope, tuff, ash, dacite,

Mt. Rainier and Mt. Lassen are both **composite volcanoes** or **stratovolcanoes** built up from successive eruptions of various types of volcanic rock, ash and other types of **pyroclastic** rock. The most abundant type of volcanic rock of stratovolcanoes is **andesite**. Mt. Rainier and Mt. Lassen are part of the Cascade Range of Mountains beginning in British Columbia, Canada and ending with Mt. Lassen in northern California.

Mt. Rainier National Park

Andesite is called an intermediate lava, which is a rock between basalt (a dark colored volcanic rock) and rhyolite (a light colored volcanic rock). In other words, it is what I call, a *blah* rock, dark to light gray often with visible sodium feldspar crystals. Its composition does include **quartz**, at 52-63% of its makeup. Since the amount of quartz can influence the explosivity of a volcanic eruption, volcanoes like Mt. Rainier and Mt. Lassen can be quite violent.

Mt. Rainier National Park

Mt. Rainier National Park was established as our country's fifth national park in 1899. It comprises 369 square miles and consists primarily of beautiful Alpine meadows and the tallest active volcano in the continental US at 14,411 feet tall. It is located in southwestern Washington in a fairly straight line with other stratovolcanoes including Mount St. Helens, Mt. Adams in Washington and Mt. Hood, in northern Oregon. Mt. Rainier is considered to be the most dangerous volcano in the continental US not just because it is an active volcano, but because of the combination of the danger from *lahars* and the 250,000 people that live around the base of the volcano. These lahars or landslides are made up of loosely compacted volcanic rock, ice and previous pyroclastic material that can be easily triggered into landslides by melting ice or earthquakes. Mt. Rainier experiences over a thousand earthquakes per year. Two volcanic craters top the summit, each more than 1,000 feet in diameter. The most recent recorded volcanic eruption was between 1820 and 1854, but many eyewitnesses reported eruptive activity in 1858, 1870, 1879, 1882 and 1894 as well.

With 26 major glaciers and 36 square miles of permanent snowfields and glaciers, Mount Rainier is the most heavily glaciated peak in the lower 48 states.

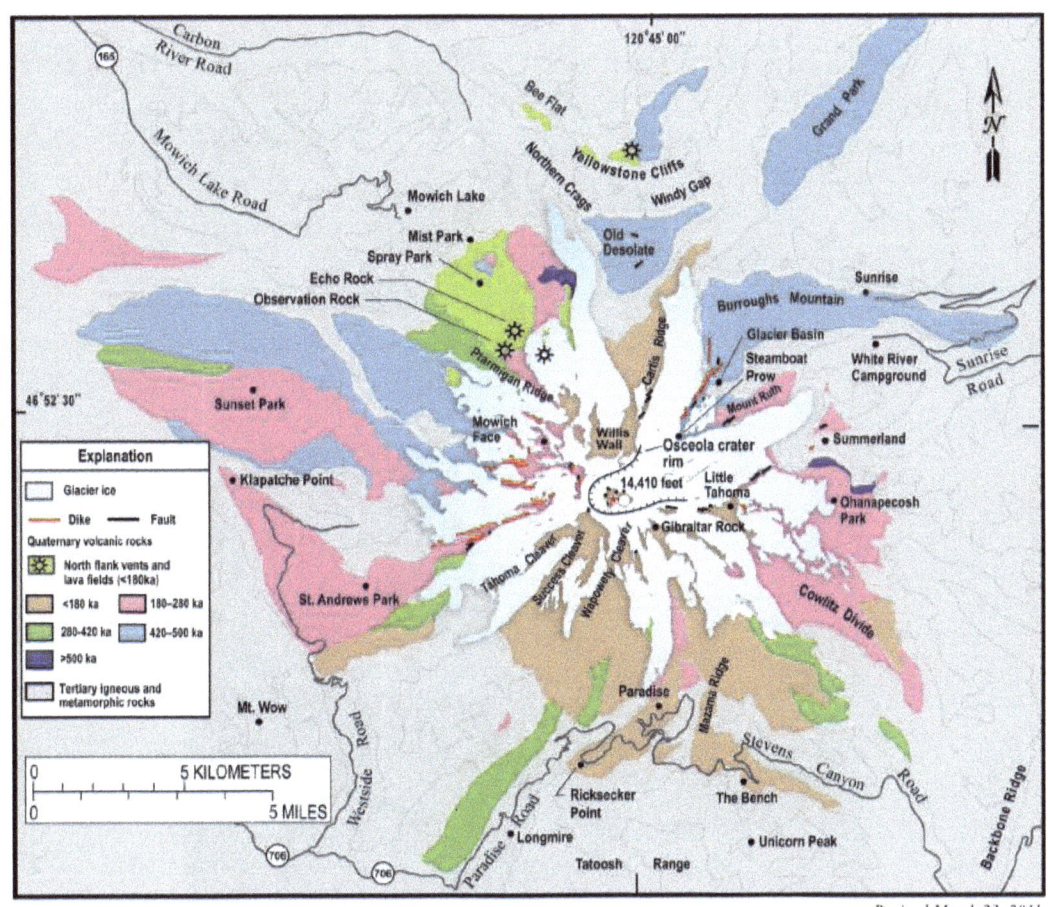

Map showing the different volcanic rocks from past eruptions

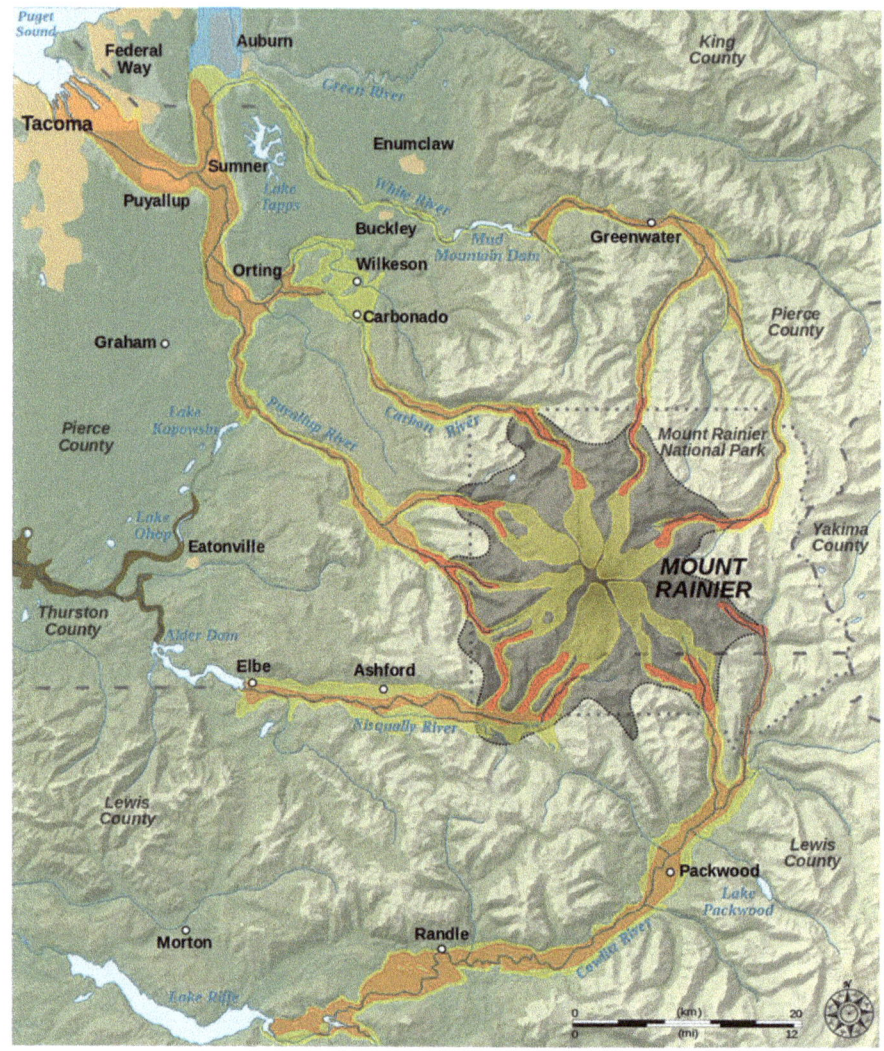

Map showing the potential lahar hazards. Five major rivers flow out of Mt. Rainier's glaciers through heavily populated communities of the greater Seattle area, including Tacoma.

Beautiful but ominous Mt. Rainier overlooking the large city of Tacoma, Washington

The rocks of Mt. Rainier are fun to explore and study. Stratovolcanoes are built of rocks from successive eruptions and these eruptions produce a variety of lavas and pyroclastic materials with unique chemical signatures.

The most abundant rock of Mt. Rainier is andesite, a light to dark gray volcanic rock that can form hexagonal columns much the same as basalt does. (Basalt is a much darker lava.)

Andesite lava   Rhyolite lava   *Tuff* (pyroclastic rock)
Rocks from Mt. Rainier

The study of Mt. Rainier is a good foundation for understanding the difference between Earth science and Earth history. All that has been written in our study on Mt. Rainier to this point is the science of Mt. Rainier. As soon as we begin to think about the *origin* of Mt. Rainier, we enter a different realm where no eyewitness or historical records have been kept.

Radiometric Dating: What is it?
Let's take a little time to discuss radiometric dating. Radiometric dating relies on the ***assumption*** that present, observed decay of radioactive elements has continued without interruption into the remote past. This cannot be scientifically demonstrated.

Radiometric Dating: A Short Explanation
Radiometric dating is based on a few assumptions about radioactivity. Radioactivity involves the decay of unstable or abnormal ***isotopes*** or elements that are slightly different from that of their normal counterparts. Radioactivity is an observed phenomenon, but no one knows for sure why it exists. At any rate, the present decay of an isotope can be measured. It is fairly constant at present. But it is assumed that it has always been constant throughout the distant past. By applying this present decay rate to volcanic rocks, which contain radioactive elements, an age can be proposed for the rock in question. This all sounds very simple and straightforward scientifically. But this assumption,

constant decay rate of radioactivity throughout the remote past, has left out one extremely important historical record – the Genesis Flood. If that global event happened the way the Bible records it did, then the geological processes of the past could have been significantly impacted to the point of invalidating all presently derived radioactive dates. All secular dates given for the geology of our national parks are predominantly based on this main assumption concerning radioactive decay. And it is this assumption that has caused a wholesale rejection of Biblical reliability and authority. Millions of people have slipped into eternity separated from God based on an assumption that has man's reasoning as its source of authority. This has been one of the main reasons that people have rejected the Bible. In my opinion, that is one of the saddest legacies of modern science.

In order to fit Mt. Rainier into a Biblical context, we have to use our Basic Biblical Framework. As *radiometric dating* relies on unproven assumptions for its usefulness, we have no way of knowing scientifically just how old this volcano is. The rocks do not come with this information, at least not without assuming certain things about them.

Secular geologists place the origin of Mt. Rainier as a small volcano at about 1,000,000 years old. The most significant volcanic activity is placed from about 500,000 years ago, to the 1800s. Our Biblical record, however, indicates an age for the Earth to be about 6,000 years old. Something is horribly wrong here. Either the secular geologists are in error or the Bible is in error. However, the Bible has already been shown to be reliable and historically accurate. So, is it just possible that secular geologists have made some serious errors in their calculations of the age of Mt. Rainier? The answer is obviously yes. All geological ages for all of our national parks are based on *radiometric dates* which all use the same set of assumptions about the behavior of radioactivity. And a little-known fact is that many of these ages have been selectively chosen to coincide with the accepted Geological Time Scale developed in the mid-1800s, long before radioactivity was even discovered. The Geological Time Scale was developed based on uniformitarian ideas about Earth's geological past. The ages and divisions of the Time Scale have been continuously revised since the 1800s. For a review of just how the radiometric dates are assigned, please see the book, *The Mythology of Modern Dating Methods*, John Woodmorappe, 1999, Institute for Creation Research.

The secular framework is built on the idea that Earth history must have been shaped by geological processes happening *small and slow and long ago.* The Biblical framework is built on the opposite idea that Earth history came about by geological processes happening *big and fast and in the recent past.* (See Lesson One.) Neither one of these ideas are scientifically testable. Although most scientists accept the former idea about the history of the Earth as scientific fact, their acceptance of this idea does not prove that the Earth is that old. It merely tells us that most scientists accept that idea but could be wrong! A choice has been forced on us, not because of science, but because of conflicting philosophies of Earth history. Either the Earth is billions of years old or it is thousands of years young. Fitting the eruptions of Mt. Rainier into a Biblical framework would tell us that this volcano was a catastrophically produced volcano probably shortly after the Flood and just prior to the great ice event known as The Ice Age. The glaciers which now cover Mt. Rainier are remnants of that ice event.

With the development of radioactive dating techniques in the early 1900s, geologists began publishing ages for the various geological landforms of our national parks. Though no one knows for sure how many attempts are made to arrive at an age of origin for various formations, we can be sure that none of the ages will line up with the Biblical interpretation of Earth history. Radiometric dating has become the cornerstone for validating secular geology and it is *the* standard of dating for all the national parks. Neither secular scientists nor most laypersons question the scientific validity of radiometric dating today.

Lassen Volcanic National Park

Lassen Volcanic National Park, often called Mt. Lassen National Park, started as two separate national monuments designated by President Theodore Roosevelt in 1907 – Cinder Cone National Monument and Lassen Peak National Monument.

View of what is left of Mt. Lassen after its 1915 eruption

Mt. Lassen is 10,457 feet high and its lava dome is one of the largest on earth at .5 cubic miles of lava rock. On May 22, 1915, a powerful explosive eruption at Lassen Peak devastated nearby areas and spread volcanic *ash* as far as 200 miles to the east. This explosion was the most powerful in a series of eruptions from 1914 through 1917. Lassen Peak and Mount St. Helens were the only two volcanoes in the contiguous United States to erupt during the 20th century. Unlike Mount St. Helens, however, Mt. Lassen exhibits a number of hydrothermal features similar to some in Yellowstone National Park.

Mud pot at Sulphur Works; the bubbles are escaping carbon dioxide.

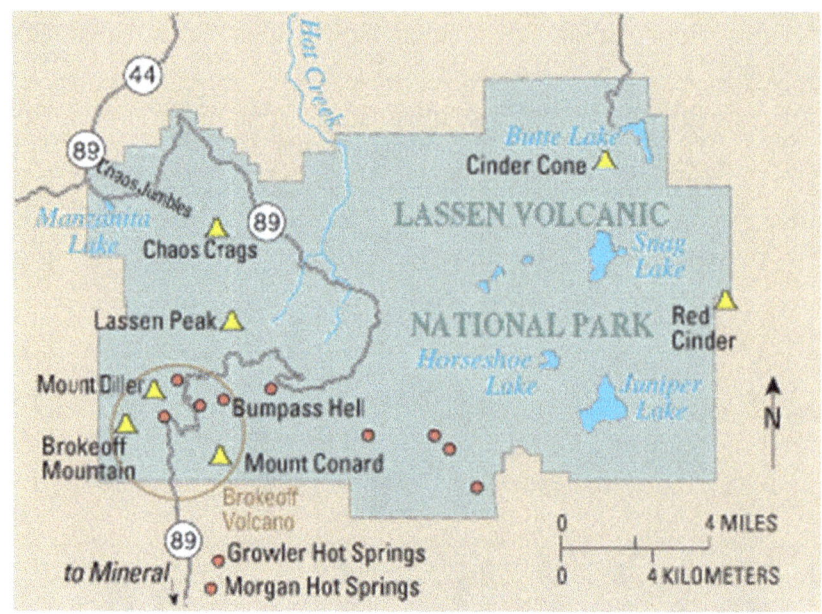

Map of Lassen Volcanic National Park showing the location of some of the hydrothermal features there (red dots)

Mt. Lassen before and after 1914

Secular geologists date Mt. Lassen at about 27,000 years old. But of course, that date is derived from the usual radiometric dating of volcanic rocks to which we have already referred. Again, radiometric dating depends on several assumptions about the past behavior of radioactivity. In order for radiometric dating to work, these assumptions must be accepted.

Mt. Lassen has four types of volcanic rocks.
- **Rhyolite lava** contains greater than 68% quartz, which makes it the result of very explosive eruptions.
- **Tuff** comes from a Latin word and is used to categorize fused or compacted ash with small bits and pieces of volcanic rocks or minerals. I am not sure why it is called *tuff*. Tuff belongs to the class of ***pyroclastic*** volcanic rocks. The word pyroclastic means *fire broken*. That word appropriately describes the makeup of

this rock. It contains broken bits and pieces of volcanic rock within a fine-grained matrix of ash.
- *Dacite* lava is an intermediate volcanic rock between andesite and rhyolite. It can look like andesite but chemically it contains between 20-60% quartz. Because of its high *viscosity*, it often leaves evidence of its flow patterns.
- *Andesite* lava is an intermediate volcanic rock between dacite and basalt. It is most often a light to dark gray-colored rock that exhibits sodium feldspar crystals. It usually has less quartz than dacite, but more than basalt. And it has less iron than basalt, but more than dacite and rhyolite.

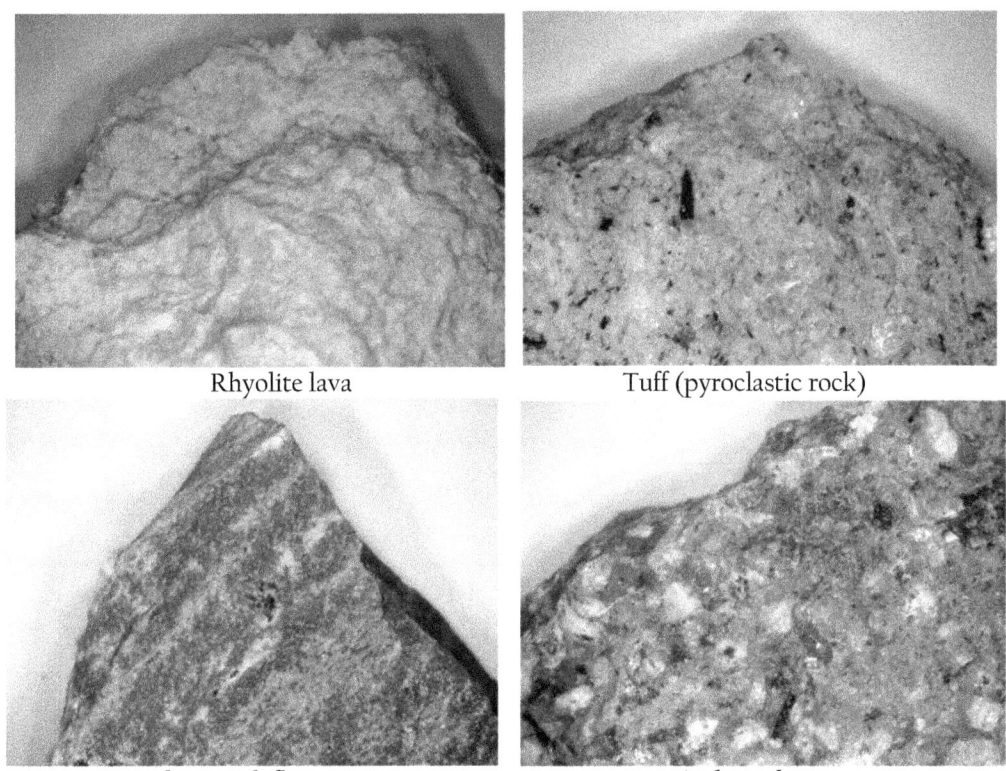

Rhyolite lava   Tuff (pyroclastic rock)

Dacite lava with flow patterns   Andesite lava

Four volcanic rock samples from Mt. Lassen

Thought Questions

1. What type of volcanos are Mt. Rainier and Mt. Lassen?

2. What is a lahar? Why is it so dangerous pertaining to Mt. Rainier?

3. What is tuff?

4. What is the difference between dacite and andesite?

5. When did Mt. Lassen erupt?

6. Discuss some of the devastation that Mt. Rainier could cause.

Activity: Illustrate the differences between a caldera and a stratovolcano.

Lesson Five – Glaciers and the Ice Age:
Yosemite National Park

Word challenges: granite, diorite, granodiorite, batholith, glacial troughs, erratic boulders, plutonic rock

Yosemite National Park

Yosemite National Park

Yosemite became a National Park in 1890. Yosemite National Park covers an area of 1,168 square miles and is located in the east central part of the State of California. Over four million people visit Yosemite each year, one million more than Yellowstone National Park! Yosemite is part of the extensive Sierra Nevada Range of Mountains, which runs 400 miles north to south, and is approximately 70 miles across east to west. Notable features of the Sierra Nevadas include Lake Tahoe, the largest alpine lake in North America; Mount Whitney at 14,505 feet, the highest point in the contiguous United States; and Yosemite Valley sculpted by glaciers out of the *plutonic rocks granite*, *diorite* and *granodiorite*. The name *Yosemite*, which means *killer* in Miwok (Native American tribe), originally referred to the name of a renegade tribe which was driven out of the area (and possibly annihilated) in the Mariposa wars.

Yosemite National Park is primarily known for its obvious glacial evidence, and massive granite formations, which are part of what is known in geology as the Sierra Nevada Batholith. The word **batholith** is from two Greek words, meaning, *depth rock*. Young Earth and Old Earth geologists agree that the large granite formations of Yosemite National Park were pushed up from deep underground. The real dispute concerns when this happened, and the amount of time it took to accomplish this geological movement. Using our Basic Biblical Framework, we would conclude that the Genesis Flood with its massive global tectonic processes implied in Genesis 7:11 and in Psalm 104:5-9, would have taken the original plutonic rock and metamorphosed it into gneiss and schist. These are both a part of the Sierra Nevada Range. Remember that our Biblical framework would teach, *big and fast and in the recent past.*

Within a few hundred years of the end of the Flood, extreme climate changes due to the prolific amount and frequency of volcanic eruptions such as Yellowstone, Mt. Mazama, and the volcanoes of the Cascade Range, would have produced significant amounts of volcanic ash, blocking the sun's reflective energy; this would have contributed to a much cooler atmosphere. That along with another significant factor, the breaking up of the fountains of the great deep would have produced warmer oceans. This would have caused increased evaporation of moisture that would have condensed in a cooler atmosphere, and then fallen back to the earth in the form of significant amounts of snow. More snow would have fallen than melted, producing significant glacial build up in a short period of time. It is this great ice event which would have sculpted the granites of Yosemite, leaving the familiar rounded rock features, ***glacial troughs*** and ***erratic boulders*** so prominent in Yosemite today.

One of the most familiar landmarks in Yosemite is Half-Dome, a granite mountain that was split by a glacier during the great ice event.

Notice in the background of this picture the striated patterns and rounded granite. This is evidence of a massive amount of moving ice.

Glacial erratic boulders in Yosemite moved by ice

Glacial trough in Yosemite carved by ice in hard granite. The lake in the foreground is a glacial lake leftover from the melting glaciers. The haze in the picture is from the famous Yosemite fires of 2014

Glaciology is the study of glaciers. The Latin word, *glacies*, means *frost* or *ice*. So, glaciology is literally the study of ice. Although there is a specialized field within geology today called glaciology, there is no unanimous agreement on just how an ice age can start or end. Over 100 ideas have been proposed for the origin of ice ages, but they usually involve too much cold weather, or not enough snow, for a long enough period of time.

It is generally accepted that Louis Agassiz, a Swiss-born scientist, is called the Father of Glaciology. Interestingly it was Agassiz who first called the evidence of glaciers, an *ice event*. He did not call it an *ice age*. The term *ice age* was introduced to reflect the belief in an old Earth. When Agassiz first presented his research for this ice event in America, it was received with quite a bit of criticism. Geologists of that time were convinced that the Earth had been born out of a molten ball of rock and it was thought to be still in the process of cooling down. An ice age just did not fit with their current framework.

Louis Agassiz, the Father of Glaciology, 1807-1873

The primary rocks of Yosemite National Park are ***plutonic rocks***. These are rocks in which one can easily see and identify the mineral crystals. The principle rocks present are granite, granodiorite and diorite. They are different from one another primarily in the amount of quartz they contain. Granite contains an abundance of quartz, close to 60%. Granodiorite contains less than 20% and diorite usually does not contain quartz or it contains less than 20%.

Granite   Granodiorite

Diorite
Plutonic rocks from the Yosemite National Park area

Thought Questions

1. Describe the word, *batholith*.

2. What is an erratic?

3. How does a glacial event explain the geologic features in Yosemite?

4. Describe the Sierra Nevada Range of Mountains.

Activity: Freeze water in a large container or obtain a large, heavy block of ice. It must be heavy for this activity to work. Put a heavy-duty piece of string around the block and drag it over a piece of wood with sand scattered over the wood. Observe what happens. Did you find any evidence that the ice had moved over the wood?

Lesson Six – Orogeny
Grand Teton National Park

Word challenges: orogeny, fault-block mountain, tectonic, metamorphic rock

What is **orogeny**? The word comes from Greek roots meaning, *mountain genesis*. It concerns itself with the study of the origin of mountains. In this regard, there is no better range to study than the Teton Range of mountains known as Grand Teton National Park.

Grand Teton National Park

Grand Teton National Park is located in northwest Wyoming just ten miles south of Yellowstone National Park. It is comprised of almost 500 square miles of breathtaking mountain peaks that have been sculpted by glaciers of the past. The Teton Range is 40 miles long. Grand Teton National Park is named for the tallest peak in the range, Grand Teton, 13,775 feet tall.

More than 1,000 species of plants, dozens of species of mammals, 300 species of birds, more than a dozen fish species and a few species of reptiles and amphibians exist within the park. It became a national park in 1929.

The classic picture of the Tetons: this background has been used for numerous western movies.

Geologically the Teton Range is called a *fault-block mountain*. Fault blocks are very large blocks of rock, sometimes hundreds of miles in extent, created by *tectonic* and localized stresses in the Earth's crust. Does this remind you of anything concerning the Earth movement involved in the Flood? The Tetons are so unusual because this block of uniform rock rises directly up from the flat plain to the east of it, the Snake River Plain. It is as if someone rapidly pushed this block of rock directly up into the air without disturbing the land immediately around it. Here is the way it looks on paper.

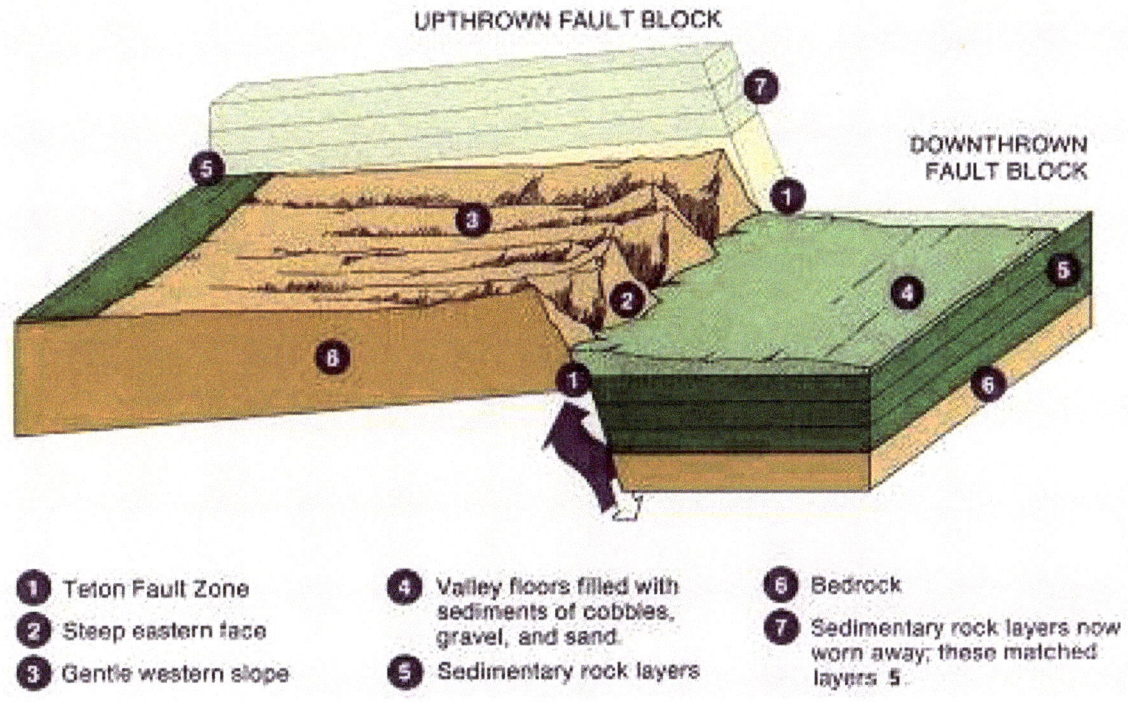

The Teton Fault Block as visualized by secular geologists

Both secular and Flood geologists are in 100% agreement as to the science of the Tetons. The real questions have to do with the *time* involved in accomplishing this amazing geological orogenic event where an up-thrown fault was pushed up over 7,000 vertical feet above the plains to the east.

Secular geologists date the rocks of the Tetons across a wide spectrum of ages from 2.6 billion to 12 million years old. The dates are derived from radiometric dating. Rocks of the Tetons include ***metamorphic rocks***, granites, volcanic rocks, sedimentary rocks and glacial sediments.

Secular geologists acknowledge that the former landforms which make up the Tetons were once under water. In fact, one of the most interesting phenomena is a sliver of sandstone at the top of Mt. Moran that contains marine fossils! Mt. Moran is one of the highest peaks in the park, at over 12,000 feet in elevation. How did this happen?

Mt. Moran is a 12,605 ft. tall metamorphic rock block with a sliver of marine sandstone at the top (in the yellow circle).

If we revisit our Basic Biblical Framework, we can see that orogeny was very much a part of the Genesis Flood. In fact, if you will read Psalm 104:5-9, you will see this same process at work toward the end of the Flood. *"He established the earth upon its foundations, so that it will not totter (the creation). You covered it with the deep as with a garment; the waters were standing above the mountains (The Flood). At Your rebuke, they (the waters) fled, at the sound of Your thunder they (the waters) hurried away. The mountains rose (orogeny); the valleys sank down to the place which You established for them."* Beginning with the onset of the Flood and continuing for a little over a year, there was significant tectonic activity. But especially as the waters began to rush off the Earth, the mechanism for accomplishing that was the orogeny of the various mountain chains brought on by the rebuke of God. *"The mountains rose; the valleys sank down...."* aptly describes the formation of the Tetons. It also explains why we find a remnant of a water-laid rock at the top of Mt. Moran of the Teton Range with marine fossils!

The rapid rise of the Tetons would also explain the metamorphic rocks that are the predominant rock of the Tetons. Rapid movement of rock would have generated lots of friction and, consequently, heat. This heat could have been responsible for transforming the basement plutonic rocks into the metamorphic rocks that now form the vistas of Grand Teton National Park.

Thought Questions

1. Define the word *orogeny*. Explain how the Biblical Framework could account for this geological event.

2. What is a metamorphic rock? How might the orogenic event of the rise of the Tetons explain the existence of metamorphic rocks?

Activity: Compile several pictures of the Tetons and find out what rocks make up each peak. Apply your Biblical Framework and attempt to explain how each rock might have formed.

Lesson Seven – Erosion, Slow and Fast:
The Five National Parks of Utah

Word challenges: sandstone, sedimentary, cross-bedding, turbidity current, hoodoo, mesa, laccolith

Three examples of sandstone that make up most of the sedimentary features of the five national parks of Utah

Sandstone located in Zion National Park

The national parks located in Utah are all part of the Colorado Plateau, 130,000 square miles of mostly the ***sedimentary*** rock, ***sandstone***.

The Colorado Plateau is a 130,000 square mile raised plateau, meaning that it has been planed by a force that remains a mystery in modern geology. The *planation* surface, as this plateau is called, seems to have been planed by the sheet erosion phase of the receding floodwaters of the Genesis Flood. See Lesson Three.

74

Sandstone is a water-born sedimentary rock made up of tiny, and in most cases, rounded quartz crystals cemented together by a combination of either quartz or silica and iron. The iron gives the sandstone its brownish to reddish color.

The five national parks of Utah include:

- **Zion National Park** is 229 square miles of canyons and *cross-bedded* sandstone formations.

Zion Canyon

Cross-bedding is a geological term that refers to sedimentary structures called *sets*, which are groups of inclined layers, and the inclined layers are known as cross strata.

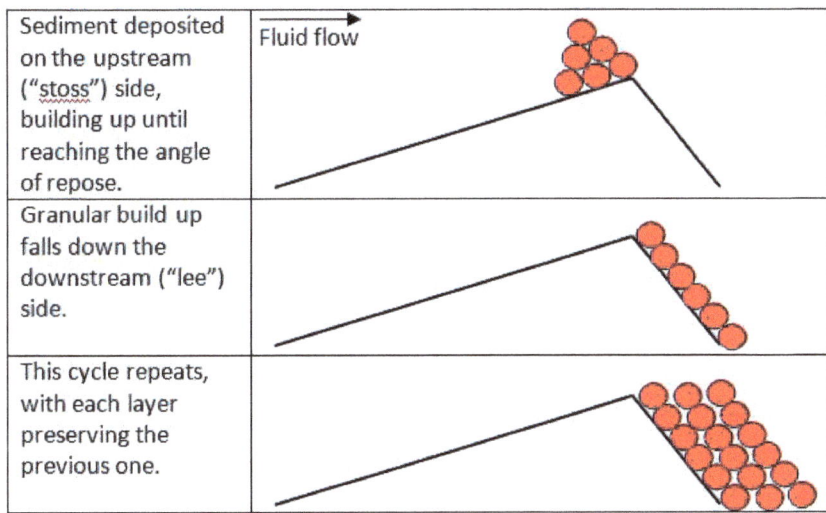

Diagram of idealized cross-bedding formation

Cross-bedded formation caused by underwater *turbidity currents*

For many years these cross-bedded structures were taught as wind-blown sand dunes and they were used as proof of the desert condition of the Colorado Plateau during the Cretaceous Period, 65-125 million years ago, during what is also referred to as the last portion of the *Age of Dinosaurs*. It has only been within the last 40-50 years that these sandstone structures have been explained as being caused by underwater currents or ***turbidity currents***. A turbidity current is an underwater current flowing swiftly downslope owing to the weight of sediment it carries. Essentially a turbidity current is a giant underwater landslide. Now, it is true that cross-bedding takes place in desert environments as wind-blown sand dunes. And some secular scientists call these formations in Zion, petrified wind-blown sand dunes. But how could they petrify if they are wind-blown? They can't, not without water!

One of the tests I have personally performed throughout the Colorado Plateau is the acid test. Wherever I have dropped a little bit of acid onto the sandstone in the Colorado Plateau, the acid has reacted by fizzing, indicating the presence of calcium carbonate. Calcium carbonate is typically associated with water-born sediments such as limestone. In the case of sandstone, the calcium carbonate provides the cementing agent in a watery environment. Could the Genesis Flood have actually produced these cross-bedded structures? The sheer size of these formations spread throughout the world would indicate yes, they could be the remnants of a global flood that covered the earth 4,500 years ago. Since then, slow erosion due to wind, rain and cold temperatures have taken their toll on the formations within Zion National Park. They have continued to erode the remnants of sandstone that were piled high in an underwater flood.

- **Capitol Reef National Park** was established in 1971 to preserve 377 square miles of sandstone features. The Park was named for a line of cliffs of white Navajo Sandstone with dome formations – like the white domes often placed on capitol buildings – that run from the Fremont River south for 60 miles. The word *reef* in this context, refers to any rocky barrier to land travel, just as ocean reefs are barriers to sea travel.

The Domes of Capitol Reef are remnants of sandstone that once covered the Colorado Plateau.

Beyond these there are two other geological features that stand out to me. The first is the prolific amount of water ripples preserved in the sandstone.

Another prominent feature not obvious to most visitors is the massive number of water-tumbled basalt boulders strewn about on the surface of the ground in Capitol Reef National Park. These volcanic rocks are not native to Capitol Reef and ice sheets did not move them there. Where did they come from? Located a few miles to the north lies the Marysvale Volcanic Field – a compact and dense compilation of a variety of volcanic features, including stratovolcanoes, calderas, lava domes, and cinder cones. Again, we are looking at a catastrophic origin for these features that would have come at the end of the Genesis Flood period as the mountains were rising up and the great ice event was rapidly building. These boulders would most likely have been moved into the Capitol Reef area as the glaciers in the volcanic mountains to the north were rapidly melting and causing major flooding.

Large water-tumbled basalt boulders – Capitol Reef National Park

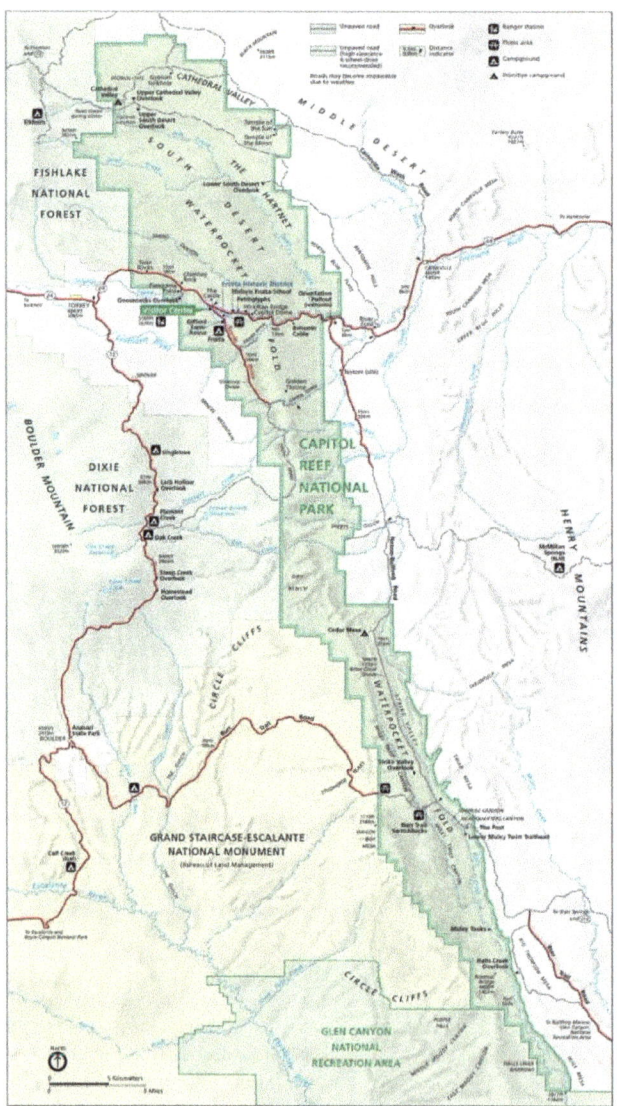

Map of Capitol Reef National Park – about 60 miles long and six miles wide

- **Bryce Canyon National Park** is comprised of 55 square miles of amphitheater-like structures of eroded sandstone. The park is located in southwestern Utah and became a national park in 1928. The so-called canyons are technically not canyons, but rapidly eroded depressions. The structures within the amphitheaters are called *hoodoos*. These are erosional remnants of the rapidly formed depression that appears to be very much like a sinkhole.

Bryce Canyon amphitheater

- **Canyonlands National Park** – located in southeastern Utah and consisting of 527 square miles of magnificent canyons and *mesas*, this area was made a national park in 1964. It is reminiscent of the Grand Canyon in Arizona. It exhibits both the sheet erosion phase of the Flood (beginning on day 151 of the Flood), and the channelized erosion phase of the Flood with its massive canyon-cutting power. The first stage of the erosion from the Flood, is called the Sheet Erosion Stage. During this stage, the water that covered the highest points of the Earth began to roll off in sheets, as the mountains rose. This event would have had the effect of shearing millions of tons of sediments, or planing the surfaces smooth, leaving extensive plateaus. Notice the flat sheared plateaus in the picture below. The second phase of the receding floodwaters would have finished draining the floodwaters in channels, leaving gashes in the freshly planed surfaces. You can see the cutting power of this draining water in the picture below. The rock that covered this area would not have been completely solid yet. It would have been cut rapidly and with little resistant. The remnants you see in the picture below would have remained.

The classic view of Canyonlands National Park:
the blue arrow points to a planed surface, and the white arrow points to channelized erosion.

- **Arches National Park** is located in southeastern Utah and literally across the road from Canyonlands National Park outside of Moab, Utah. It consists of 119 square miles of wonderfully carved sandstone arches. Many of these have broken and fallen within the last 40 years. There are over 2,000 arches in Arches National Park!

The icon of Arches National Park: Delicate Arch.
It is accessible by a short hike from the parking area.

There are two opposing views of how these arches originally formed. One is that they were carved by wind over millions of years. The other is that they were originally remnants of sandstone from the Genesis Flood. Both receding and channelized floodwaters would then have played a part in initially carving these wonderful structures. After the Flood, on-going erosion would have taken over and continued to weather these weak landforms.

And this erosion is continuing even today. No wonder Arches National Park is losing its arches at an alarming rate.

One of the most overlooked features that lies in the background of both Canyonlands and Arches National Parks are the La Sal Mountains, east of Moab, Utah. These are no ordinary mountains. They are what geologists refer to as *laccoliths*. The word *laccolith* means, *pond stone*. Laccoliths are magmatic intrusions, which means magma was rising into the sediments above it. The La Sal Mountains are magmatic intrusions that never quite erupted into volcanoes but ponded beneath many layers of sandstone. They have survived as remnants of these intrusions. So, how did they become exposed? If we refer to our Biblical framework, these magmatic intrusions would most likely have been "ponding" as the Genesis Flood buried them under miles of sedimentary sand, which later became sandstone. Toward the end of the Flood, Psalm 104:5-9 tells us that the mountains rose up and the valleys sank down. The volcanoes that never were, became exposed as they were pushed up through the still-wet sandstone into the mountains now known as the La Sal mountains. The true nature of these mountains, that they are laccoliths, is hidden due to their massive height – as much as 12,000 feet. But they are quickly identified by the magmatic rock that makes them.

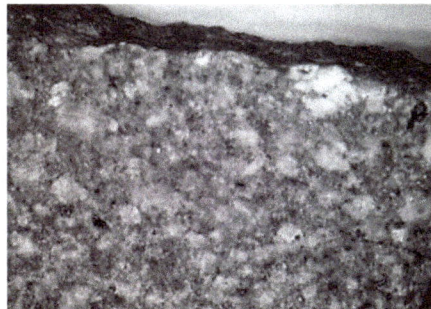 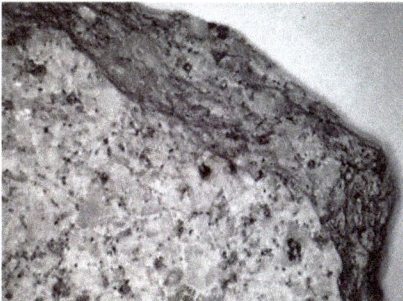

The magmatic rock of the La Sal Mountains

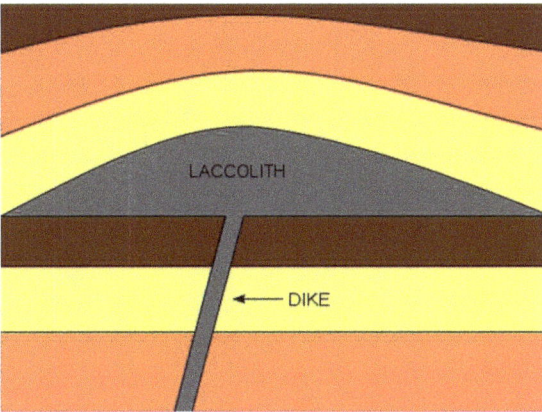

Diagram of a laccolith buried under layers of sedimentary rock above it

Round Mountain in the right of the picture, is a laccolith located just below the La Sal Mountains, themselves laccoliths. Round Mountain is surrounded by a sandstone canyon on either side, cut away as a huge channel, leaving Round Mountain as a remnant, while the floodwaters drained.

Thought Questions

1. Describe the sedimentary rock, sandstone.

2. Describe the geologic features known as cross-beds. How are they formed?

3. Give one identifying characteristic of each of the five national parks of Utah.

4. What part did the Genesis Flood play in the mystery of the rounded basalt boulders found at Capital Reef National Park.

5. What is a laccolith? How do they seem to have formed?

Activity: Obtain a microscope or powerful magnifying glass and look at a piece of sandstone with it. Next, scrape off some of the sand from the sandstone and again look at it under a microscope. Describe what you see.

Lesson Eight – Calderas and Rhyolite
Crater Lake National Park and Newberry Volcano/Caldera National Monument

Word challenges: cinder cone, shield volcano, scoria, pumice, viscous (viscosity)

Crater Lake with a *cinder cone* off to the upper left in the crater

Crater Lake National Park

Crater Lake National Park, located in southern Oregon, became a national park in 1902. It is the country's fifth oldest national park. The park encompasses a huge caldera crater of what is left of a massive volcanic eruption known as Mt. Mazama. The crater is 1,949 feet deep and is now filled with water from past glaciers of the massive ice event (or Ice Age, see Lesson Five) that followed the Genesis Flood. According to secular dating methods, the last eruption occurred about 5000 years ago. But if that is true, that would have taken place before the Flood. According to our Biblical record, however, Mt. Mazama was most likely part of the conflagration of post-Flood volcanoes that precipitated a massive ice event we now call an Ice Age. The eruption of Mt. Mazama deposited ash as far east as the northwest corner of what is now Yellowstone National Park, as far south as central Nevada, and as far north as southern British Columbia. It produced more than 150 times as much ash as the May 18, 1980 eruption of Mount St. Helens.

Crater Lake is part of the Cascade Range of mountains that contains several volcanoes of different varieties including stratovolcanoes, *cinder cones*, calderas, lava flows and *shield volcanoes*.

The Pinnacles near Crater Lake showing pale pumice flow beneath smoke-gray *scoria* flow, above which lie 10 feet of fine ash

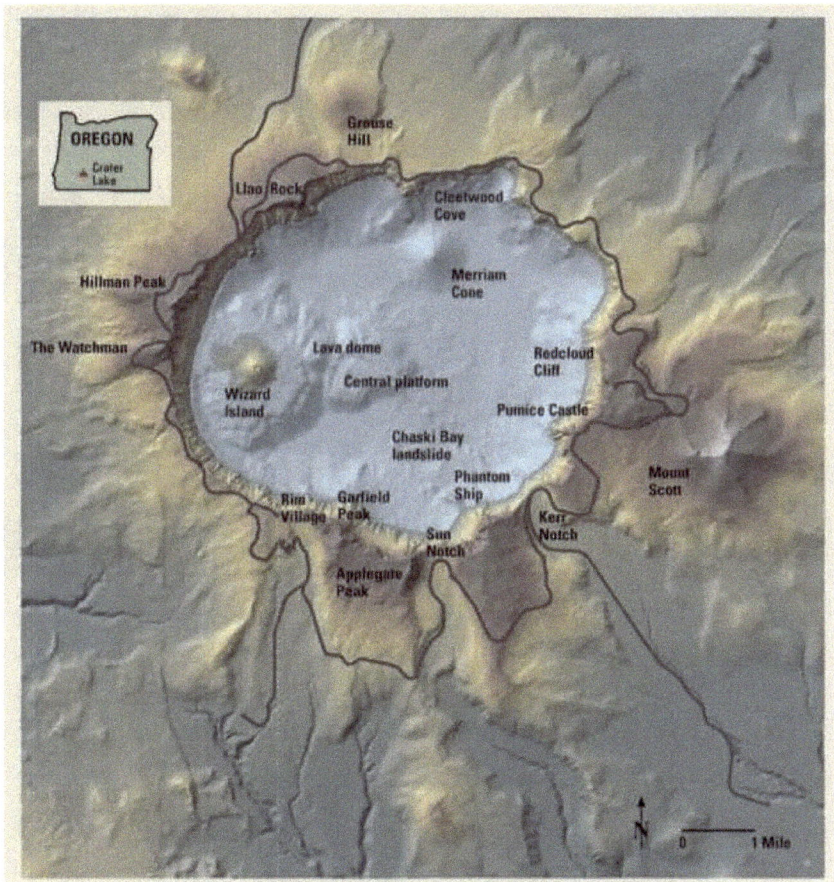

Relief map of Crater Lake

Volcanic rocks of Mt. Mazama include mostly pyroclastic rocks, as it was an explosive caldera volcano. The main pyroclastic rocks are: *tuff,* which is fused ash with bits and pieces of volcanic rocks; ash *tuff,* which is fused ash; *scoria,* which is airy basalt, "popped," typically from a cinder cone; and *pumice,* which is a glassy rock that looks like cotton candy and typical of explosive eruptions.

The most interesting drive to Crater Lake is down US Highway 97 over what is called The Volcanic Legacy Scenic Byway, a scenic and instructive drive through volcano country consisting of The Three Sisters (stratovolcanoes), Newberry Crater (shield volcano/caldera), and Devil's Lake (volcanic field) among many others. On your drive down, keep your eyes open for large chunks of pumice all along the highway. As Mt. Mazama was a huge caldera eruption, the type of pumice would have been rhyolitic with lots of quartz. Scattered along the highway, it looks like dirty cotton candy.

Volcanic rocks of Mt. Mazama: Tuff – a pyroclastic rock;   Pumice – a pyroclastic rock

Vesicular basalt – gaseous lava flow   Ash tuff – a pyroclastic rock

Scoria – a basaltic, gaseous, pyroclastic rock

If you research a history of the intensity of all known volcanic eruptions, you will notice that the most violent eruptions have all occurred in the past. Not only is the *intensity* decreasing, but the *frequency* of violent eruptions has been decreasing. Mt. Mazama would have been one of those intense eruptions of the past. The eruptions of Yellowstone would have been others. Considering the Basic Biblical Framework that positions the Flood as the primary mechanism for the origin of volcanoes, this makes sense. As the Earth has been settling down from the historical and violent ordeal of the Genesis Flood, it would seem logical that volcanic eruptive patterns would also be settling down.

Volcanic Explosivity Index and the Flow of History

Wah Wah Springs
30 Million years ago
>5500 cu km (VEI 8)

Toba
74,000 years ago
2800 cu km (VEI 8)

Yellowstone
640,000 years ago
1000 cu km (VEI 8)

Long Valley Caldera
780,000 years ago
580 cu km (VEI 7)

Krakatau
1883
20 cu km (VEI 6)

Pinatubo
1991
5 cu km (VEI 5)

Rainier
250 BC
0.30 cu km (VEI 4)

St. Helens
1980
0.25 cu km (VEI 4)

Crater Lake
7,600 years ago
150 cu km (VEI 7)

Novarupta
1912
13 cu km (VEI 6)

Vesuvius
AD 79
3.3 cu km (VEI 5?)

Eyjafjallajokull
2010
0.30 cu km (VEI 4)

All statistics are available from the USGS. Please note that the dates given for eruptions, with the exception of those that have been recorded in history, are based on uniformitarian assumptions. But even given these, it is clear that the volcanic explosivity index (VEI) and size have decreased over time.

Judging from the geological evidence, caldera eruptions would be frightening volcanic episodes. For example, geologists have predicted that if the Yellowstone caldera were to erupt again, it would take out over 250 square miles of living things!

Geologists think that a caldera eruption would begin with massive explosions of ash, pumice and other pyroclastic rocks. The crater for this would incorporate many square miles and, after the initial explosion, would then release huge amounts of rhyolitic lava rich in quartz and potassium feldspar. Due to the sheer size of the crater, the volcano would collapse in on itself leaving a huge drained depression from which lavas of various types would continue to extrude for quite some time.

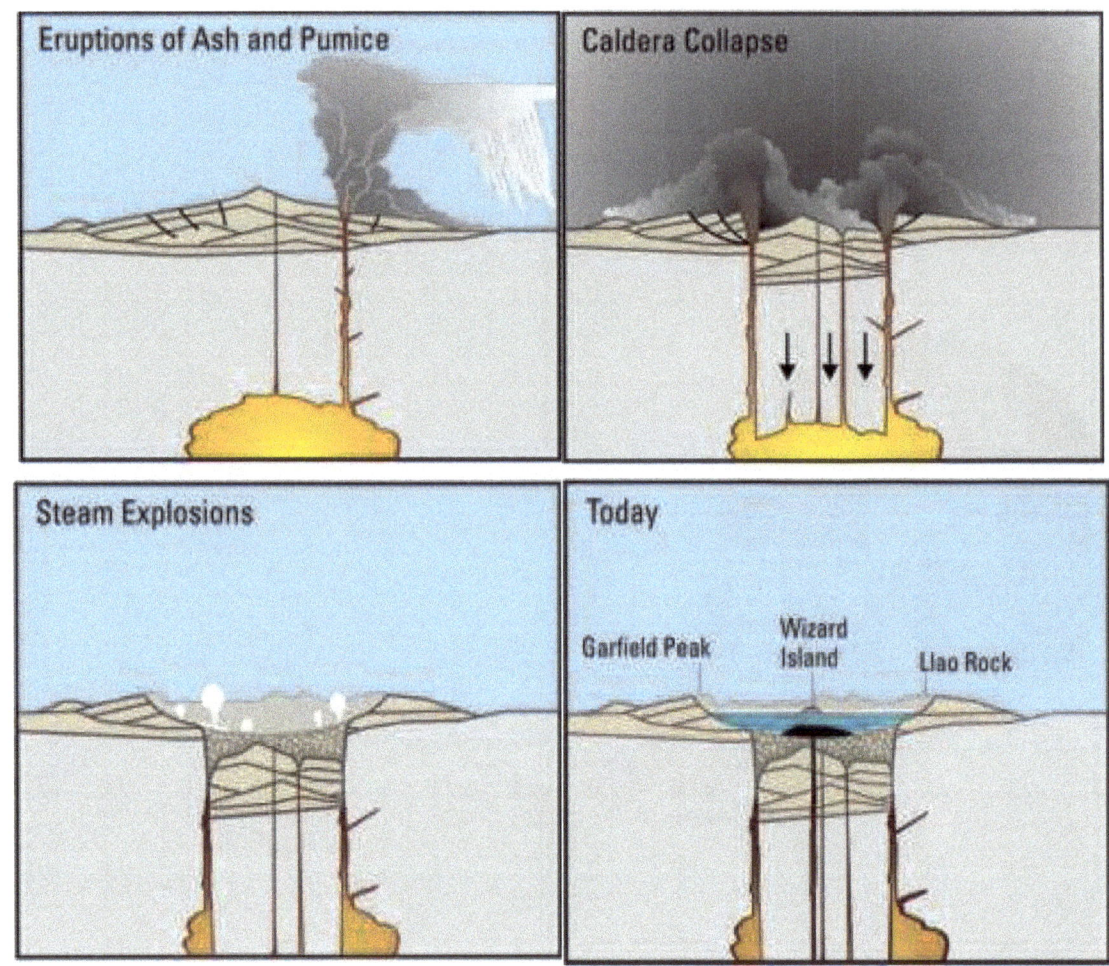

It is thought that a caldera eruption would act like the pictures above, Crater Lake being the example.

One of the reasons for the explosive nature of a caldera eruption is because of the amount of quartz involved in the lavas that are erupted. The quartz makes the lava *viscous*, or resistant to flow. And caldera eruptions are characterized by viscous lava.

Not all calderas are explosive. Kilauea, located on the flank of Mauna Loa, a shield volcano, is a huge basalt volcano, whose lava is less viscous, and so is much more fluid and less resistant to flow. Remember, viscosity is influenced by the amount of quartz lava contains. Basalt lava contains much less quartz than rhyolite, so it is not as explosive. Most of the historical caldera eruptions that are evidenced by the rocks they contain (predominantly rhyolite), were evidently quite explosive and destructive. The circumference of Crater Lake is about 33 miles.

Crater Lake, so named for the crater that is filled with glacial water; the basalt cinder cone, Wizard Island, found in Crater Lake, is in the lower left section of the crater.

Newberry Volcano/Caldera National Monument

Word challenges: cast, obsidian, rift zone,

Newberry National Volcanic Monument was designated on November 5, 1990, to protect the area around the Newberry Volcano/Caldera located in central Oregon, about 20 miles southeast of Bend, Oregon, and 75 miles northeast of Crater Lake. It comprises 78 square miles of volcanic formations including Lava Butte, Lava River Cave, Lava Cast Forest, and Newberry Caldera.

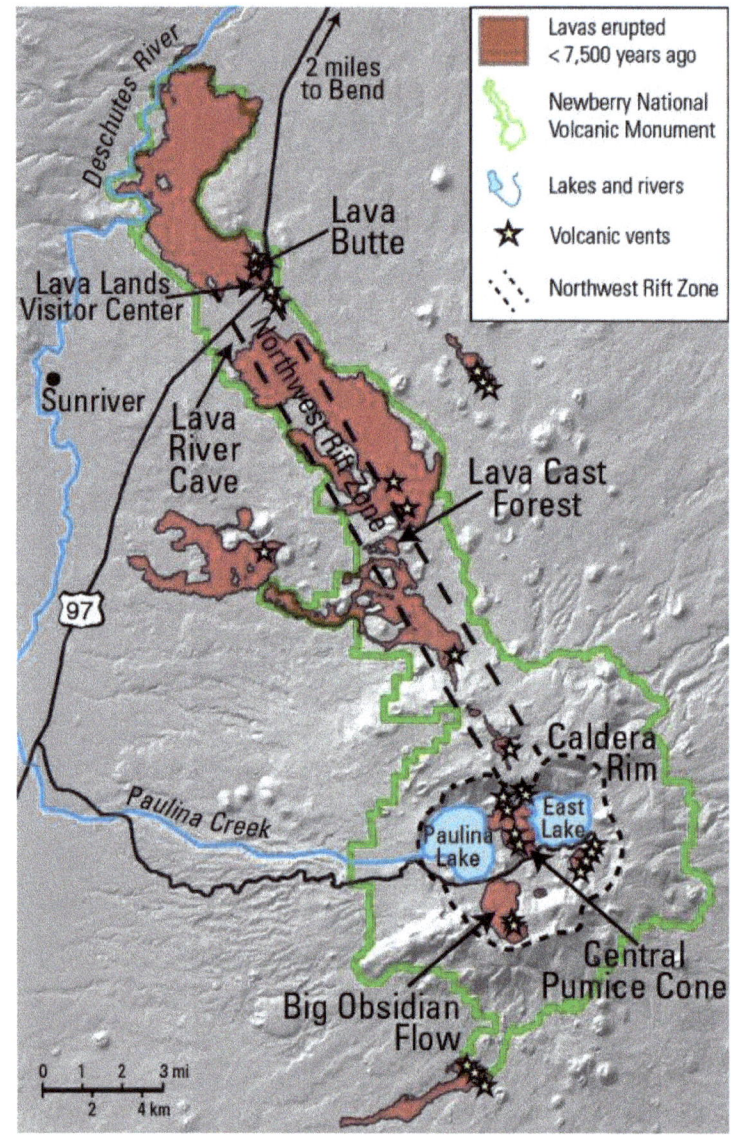

Map of the Newberry Volcano/Caldera National Monument; note the different volcanic events that took place.

Newberry National Volcanic Monument: a giant obsidian flow can be seen in the picture to the right of the lakes.

Newberry Crater for which the Monument is named is a large caldera and shield volcano. An obsidian flow is located to the right of the lake.

Newberry Caldera and Volcano are interesting features in that they contain a variety of lavas including basaltic, andesitic and rhyolitic lava flows. The rhyolitic lava flows are in the form of *obsidian*, a vari-colored lava made of pure quartz colored by varying amounts of iron.

The extent of the obsidian lava flow at Newberry Volcano/Caldera

 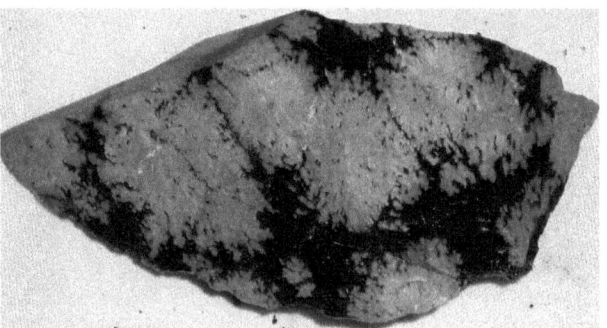

Different kinds of obsidian, a type of rhyolite

Lava River Cave, the longest continuous lava tube in Oregon at 5,211 feet long

The caldera itself is four miles by five miles. It is estimated that Newberry was around 1,000 feet higher before the caldera eruptive event. As is typical for caldera eruptions, after the initial eruption, more lava flows and volcanic events followed, as indicated on the earlier map.

How old is the Newberry Volcano/Caldera? Secular geologists date the area between 7,000 and 1,300 years old via radiometric dating. This is a little odd, especially when the accuracy of radiometric dates is so widely touted. If we assume a different framework, the age of these eruptive events would be around the same age as Crater Lake – shortly after the Flood and most likely before the great ice event. Remember, Crater Lake, in the same vicinity, was filled in by melting glaciers.

One other geological event to notice for this area is the Northwest Rift Zone, (map on p. 90) running through the Newberry Volcano/Caldera National Monument. A *rift* is a long crack in the Earth's crust. The Northwest Rift Zone is several miles long, but it is part of a much broader set of rifts, which includes the Brothers Fault Zone. These rifts dominate the geological structure of most of Oregon, some of California, Nevada, and Washington. Rifts are associated with the volcanic activity of much of the Northwest, including the caldera eruptions of Yellowstone.

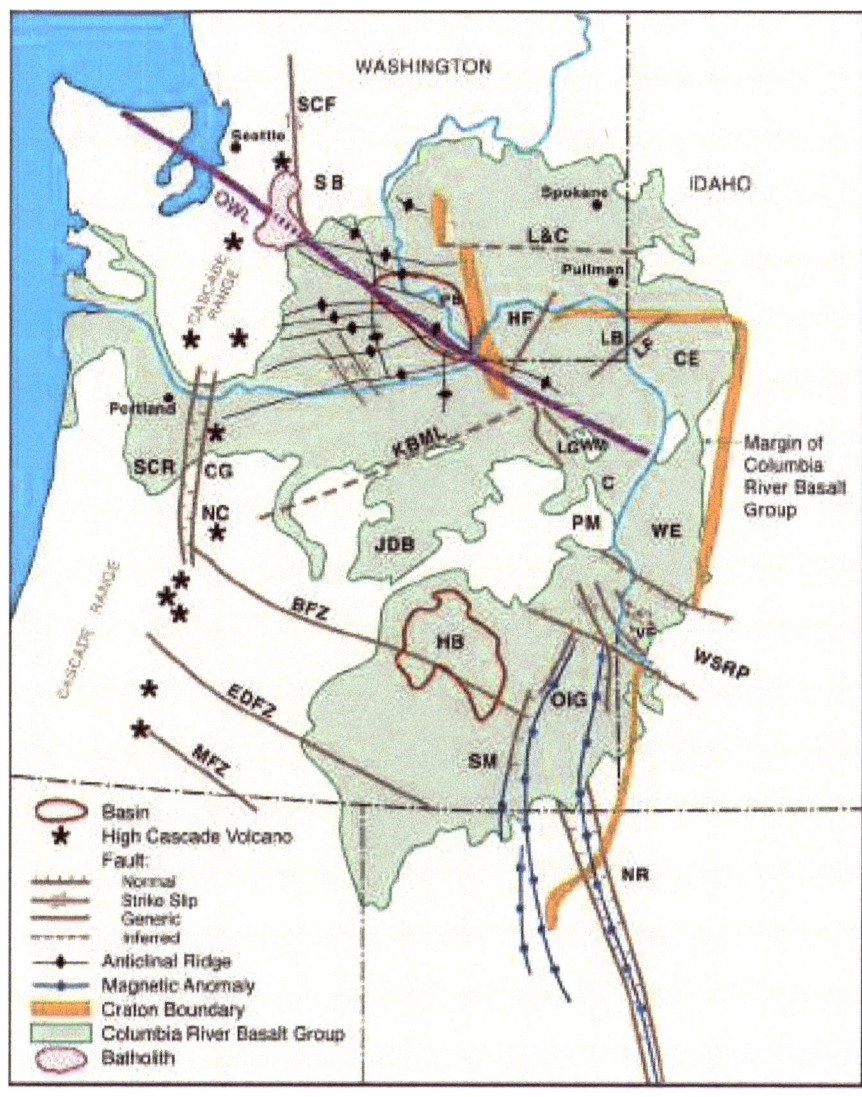

The Brothers Fault Zone or BFZ

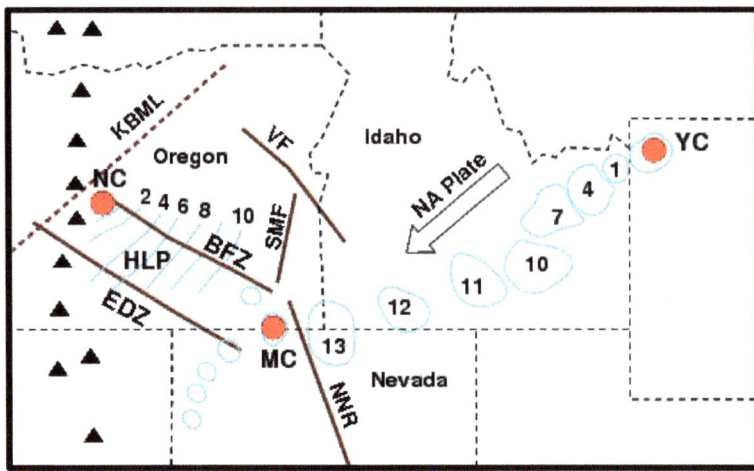

Map locating the major rifts in the Northwest and their association with the eight caldera eruptions of Yellowstone (#13, 12, 11, 10, 7, 4, 1, and YC) across the Snake River Plain of Idaho.

In summary, rifts are associated with cracks in the Earth's crust from which lava flows and eruptions come. The cause of these rifts is a matter of continuing debate. Secular geologists view them as part of the evolutionary geology of the Earth, initiated 4.6 billion years ago when the Earth started to form from a nebulous gas cloud. Within a Biblical framework rifts are about 4,500 years old and initiated by the *breaking up of the fountains of the great deep* on the first day of the global Flood of Genesis.

Thought Questions

1. Describe each of the following: cinder cone, shield volcano, caldera, stratovolcano.

2. What is the difference between pumice and scoria?

3. Why was the name Mt. Mazama changed to Crater Lake?

4. What are the characteristics of a caldera eruption?

Activity: Illustrate each of the following: A cinder cone, a shield volcano, a caldera, a stratovolcano, a lava field.

Lesson Nine – Basalt Behavior
Hawai'i Volcanoes National Park, Craters of the Moon National Monument, Capulin Volcano National Monument, El Malpais National Monument, and Lava Beds National Monument

Word challenges: volcanic vent, basalt, archipelago, pahoehoe

Hawai'i Volcanoes National Park
Hawai'i Volcanoes National Park is located on the Big Island of Hawai'i in the southeast corner of the Island. It was established as a national park in 1916. It includes an area of 505 square miles of diverse environments that include tropical forests, active volcanic lava flows, and even an arid climate in barren Ka'u Desert. The altitude covers tropical sea level to the frigid summit of the Earth's most massive active volcano, Mauna Loa at 13,677 feet.

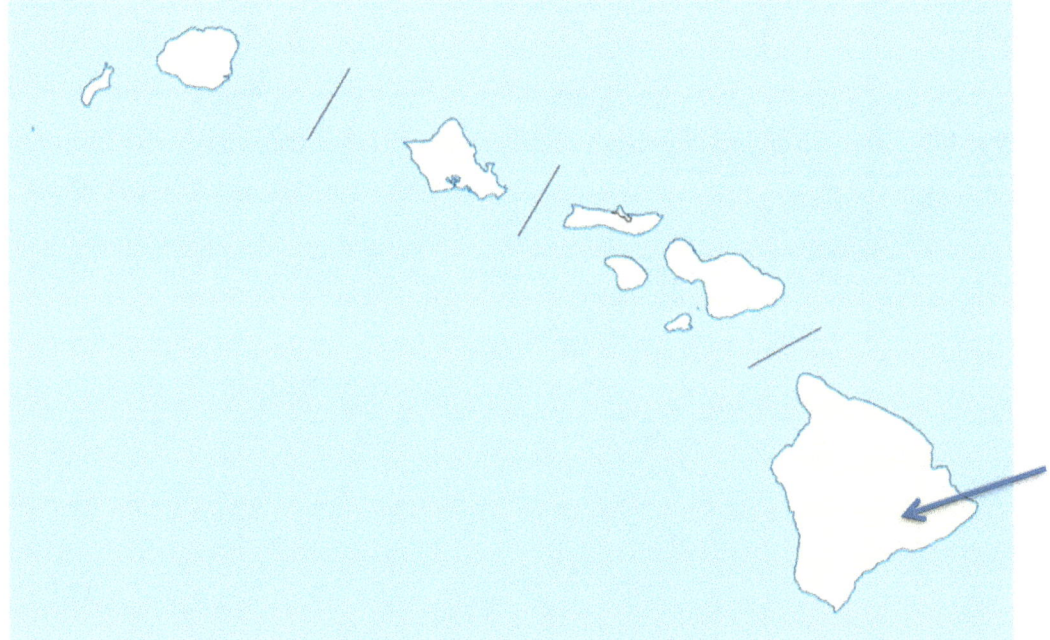

The Big Island of Hawai'i showing the location of the park

The Big Island of Hawai'i is a composite of five shield volcanoes, Kohala, Mauna Kea, Hualalai, Mauna Loa and Kilauea. The Hawaiian word *mauna* means mountain. Driving over the center of the island is quite an interesting adventure, as the trip crosses several lava flows where the highway has been destroyed and rebuilt several times.

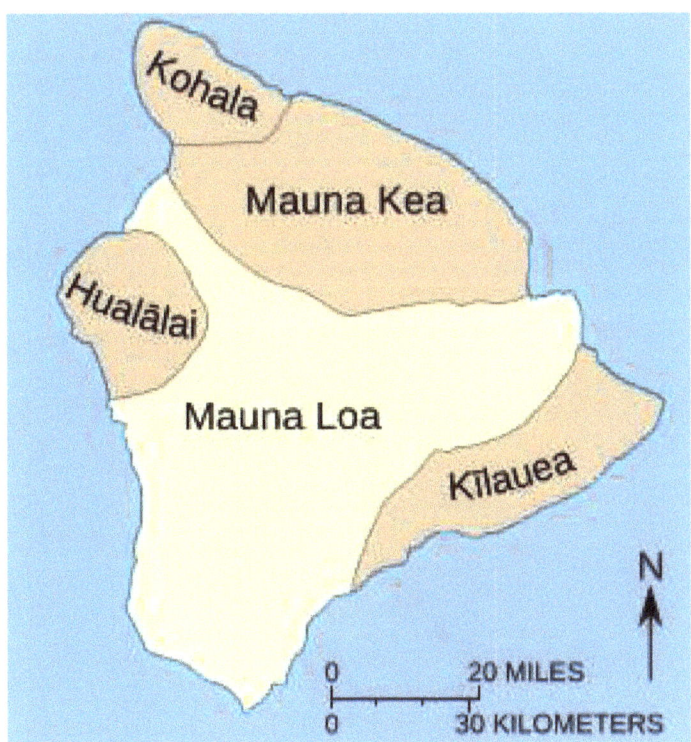
The Big Island of Hawai'i is made of five shield volcanoes.

The shield volcano, Mauna Kea

The park encompasses two active volcanoes: Kilauea, one of the world's most active volcanoes, and Mauna Loa, the world's most massive shield volcano. Over half of the park is designated as Hawai'i Volcanoes Wilderness.

The following map shows the recorded historic lava flows of the shield volcano, Mauna Loa.

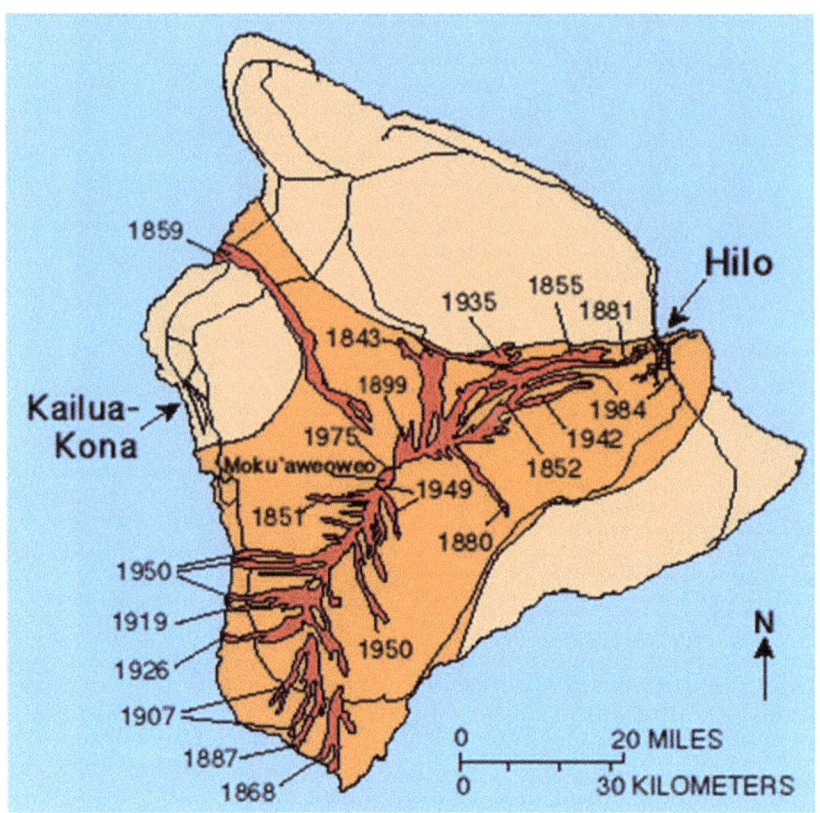

Mauna Loa is located in the center of the Big Island with historic lava flows branching out in several directions.

Active eruptive sites include the main caldera of Kilauea and a more active but remote *volcanic vent* called Pu'u'O'o.

Kilauea active crater with Mauna Loa in the background

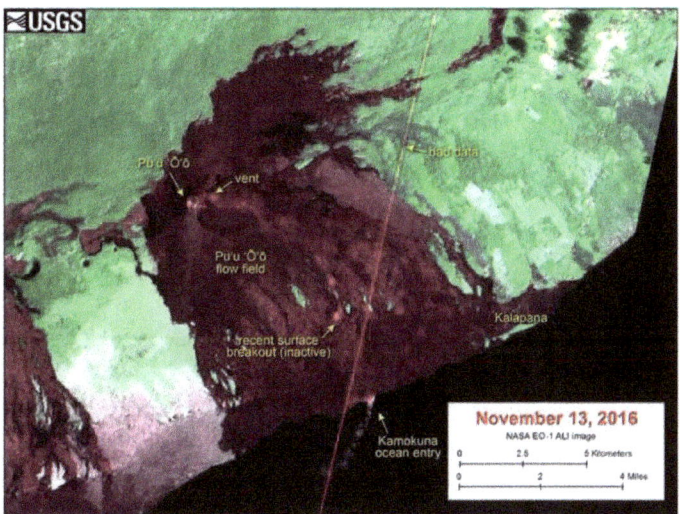

Satellite image of the Pu'u'O'o volcanic field in the southeast corner of the Big Island

Volcanic cinder cone of Pu'u'O'o

Aerial view of Pu'u'O'o taken on August 30, 1990

Lava from the Pu'u'O'o *cinder cone* has flowed 14 miles into the town of Pahoa. The lava has breached the boundary of the Pahoa Transfer Station.

March 2015, looking south toward the town of Pahoa, and southwest toward the lava flow from Pu'u'O'o, outlined by burnt vegetation

The Hawaiian Island Chain of volcanoes called an *archipelago* – satellite view

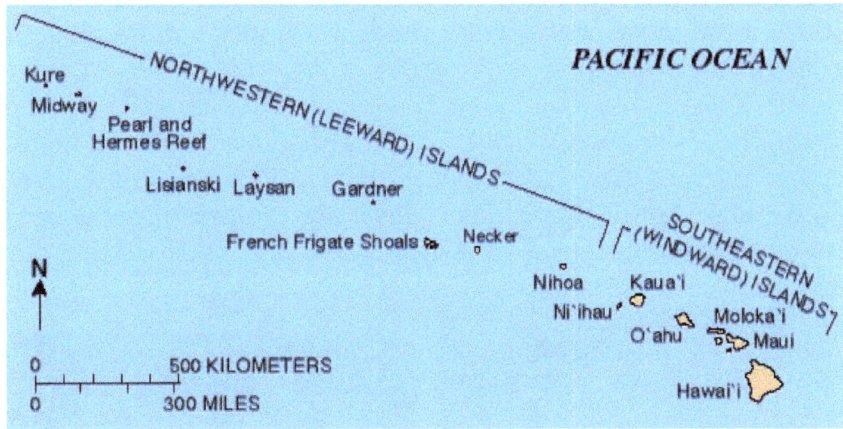

Actually, the Hawaiian Islands proper are just a small part of the larger chain of volcanic islands that stretch for over 1,700 miles.

Secular geologists date the Hawaiian Islands at over 28 million years old for the farthest northwest island and about 400,000 years old for the island of Hawai'i. These dates have been arrived at using the popular potassium/argon dating method where the parent radioactive element of potassium is supposed to have decayed over many years into the stable daughter element argon. There are two reasons why these dates are suspect. One has to do with a historical record, the Scriptures, that have been shown to be historically accurate. Therefore, the radiometric record is simply wrong. The second reason is that radiometric dates are arrived at using a major assumption. That assumption is that present radioactive rates of decay have remained constant throughout the remote past. Of course, there is no way that this can be scientifically validated. It is a guess based on the uniformitarian assumption, *the present is the key to the past.*

Also, in radiometric dating there are two other assumptions that the derived ages are dependent on. The first is the assumption that the amount of the parent radioactive element potassium *has not changed* since the time the rock was formed. The second is the assumption that the amount of the daughter stable element argon is *wholly derived* from the decay of potassium. Neither one of these can be scientifically validated. I used to teach a Proverb to my kids when they were young. *Through presumption comes nothing but strife.* Assuming things that one does not know for sure brings about strife. And that is exactly what has happened in radiometric dating. By assuming things to be true that cannot be known, the result has been strife between the science world and the Biblical worldview.

Using our Biblical framework, it is most likely that these volcanic islands did not exist prior to 2,500 BC when the fountains of the great deep broke open and released a torrent of magma and lava. Later as the ice from the great ice event melted, the sea level rose and increasing ocean water isolated many of these islands.

The Rocks of Hawai'i
The Hawaiian Islands are of course, entirely volcanic, derived from the volcanoes that make them up. The lava that comes from these volcanoes is **basalt**, which is black when fresh, high in the dark colored rock-forming minerals, olivine, pyroxene, biotite, amphibole, calcium feldspar, and iron.

Pahoehoe is ropy shaped basalt lava. The Hawaiian word means, *smooth* lava. Pahoehoe seems to be due to the movement of very fluid lava under a congealing surface crust. Pahoehoe can also forms lava tubes.

Although not totally understood, with increasing distance from the source, pahoehoe flows may change into *a'a* flows in response to heat loss and consequent increase in viscosity. A'a is rough, spiky shaped lava. Both are prominent lavas in Hawai'i. The temperature of flowing basalt ranges from 1,292° to 2,192° F, depending on the distance from the vent.

Top row, left to right: vesicular basalt (oxidized because of the iron content), *pahoehoe* lava, recent basalt flow of Mauna Loa, **Bottom row, left to right:** basalt with olivine, *a'a* basalt lava flow

One of the prominent features of the Hawaiian Islands is the deeply eroded volcanic mountains. These deeply eroded *V-shaped* ravines have been cut by water. As they are V cut, they were probably cut recently and rapidly. But it is not water-run-off from years of rain. The mountains are made of very hard basalt lava. It is highly unlikely that normal rain would have made much of an impact on these hard basalt mountains. In fact, the word *basalt* means *hard*. It is one of the hardest rocks I have broken. As no one saw what transpired when these mountains were formed, using our Biblical framework, they cannot be that old. Working within the Biblical framework, a reasonable idea would be that these mountains were carved by water that covered them during or shortly after formation, most likely during the receding stage of the Flood about 4,500 years ago.

The volcanic mountains of Oahu show a tremendous amount of erosion.

Rounded boulders are another prominent feature. The rounded basalt boulders indicate that a high velocity of water current was involved in rapidly eroding and transporting them.

Water-tumbled basalt boulders, on the Big Island of Hawai'i

Lava tubes are formed when the outer crust of a lava flow cools and hardens faster than the inside of the flow, leaving a hollow tube when finished.

Inside a lava tube on the Big Island of Hawai'i

Craters of the Moon National Monument

Craters of the Moon National Monument was established in 1924 and covers 1,117 square miles. It covers three major basalt lava fields, which lie along the Great Rift of Idaho. The Great Rift extends across almost the entire Snake River Plain. It is the largest recent lava field in the contiguous US. A rift is essentially a huge crack in the Earth. Craters of the Moon contains the deepest known crack or rift at 800 feet deep, over 60 solidified lava flows, and 25 volcanic cones.

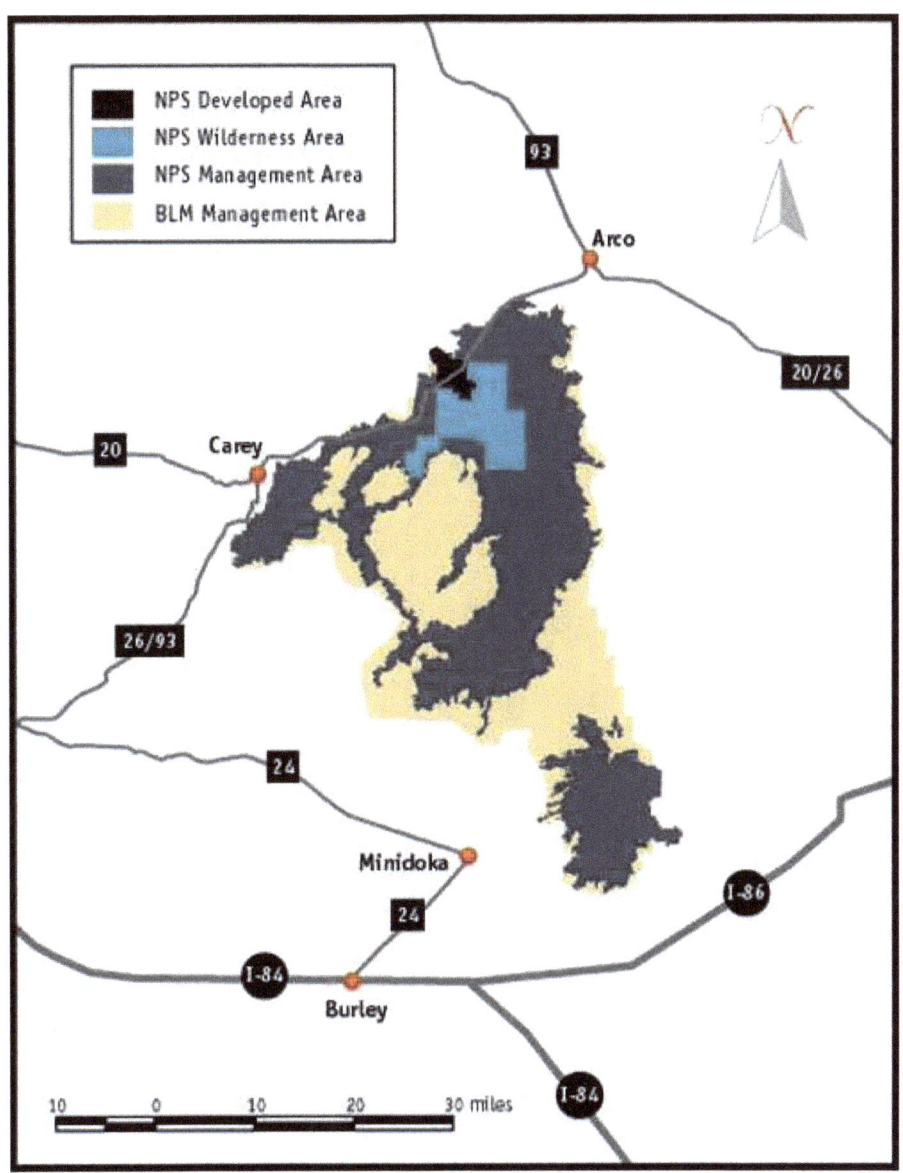

Map of the Craters of the Moon National Monument east of Boise, Idaho

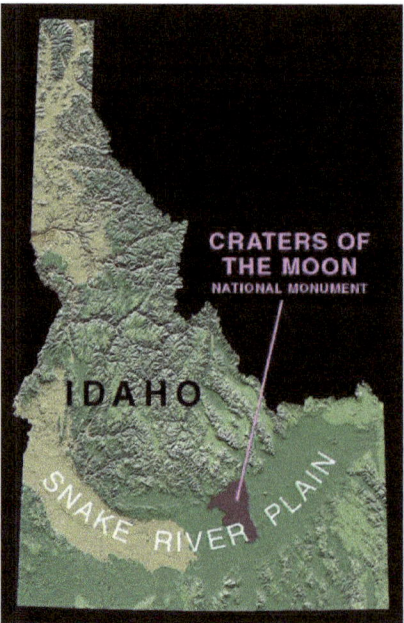

Situated along the great volcanic Snake River Plain, Craters of the Moon is part of a much larger volcanic picture stretching all the way into northwestern Wyoming and the calderas of Yellowstone National Park.

Tree mold showing bark and mold of a tree overwhelmed by lava flows

Author standing in an a'a lava field

Cinder cone of Craters of the Moon

Basalt cinders, Craters of the Moon

Pahoehoe (ropey) lava flow, Craters of the Moon

Capulin Volcano

Capulin Volcano in far northeastern New Mexico, designated as a National Monument in 1916, is a cinder cone and part of the 8,000-square mile Raton-Clayton Volcanic Field. A volcanic field is an area of the Earth's crust that is prone to localized volcanic activity. These fields usually contain 10 to 100 volcanoes such as cinder cones, usually arranged in clusters. The Raton-Clayton Volcanic Field contains multiple cinder cones, domes, tuff rings, and the immense andesite-shield volcano, Sierra Grande. Lava flows may also occur. Volcanic fields are thought to be tied into the same magma source.

Capulin Volcano is a rather large cinder cone measuring about a mile in circumference with a crater about 400 feet deep.

Aerial view showing part of the Raton-Clayton Volcanic Field

Capulin Volcano - a cinder cone

The Capulin Volcano sits astride the Rio Grande Rift, an elongated valley of rifting that extends in a north-south direction from Colorado to central Mexico. A rift is essentially a crack in the Earth's crust.

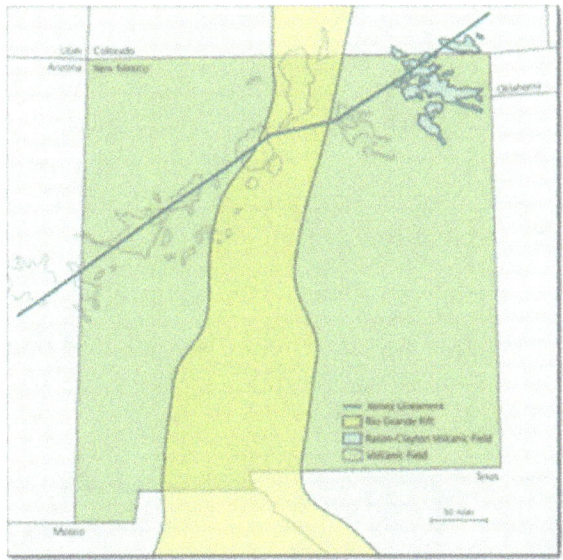

This map shows the continental rift, called the Rio Grande Rift. Notice the Capulin Volcano area in the northeast, and the rift running down the center of the state.

The fragility of the Earth's crust today is a result of either long ages of unstable geological activity since the Earth began about 4.6 billion years ago or it is the result of one massive geological event about 4,500 years ago because of the Genesis Flood, when the fountains of the great deep burst open. Neither of these can be proven to be true based on scientific investigation. Science deals with events in the present. The rifting of the crust of the Earth was a result of past geological events. It is the Biblical framework that we apply to get *a big and fast and in the recent past* interpretation.

Geologists date this volcano at around 50,000 years old because of radiometric dating. The Biblical historical framework would place it shortly after the Genesis Flood. In the photo below, one can see the *planation* surface in the background. Planation surfaces are mysteries in modern geology. They are flat erosional remnants surrounding a scoured-out valley or canyon. The immensity of many of these landforms demands a catastrophic explanation. The Genesis Flood provides both the shearing, or planing, and the channelization, or sculpting, of the valley or canyon.

This is a view of the area surrounding Capulin from the summit; notice the planation surface in the background.

El Malpais

El Malpais National Monument is a collection of lava flows, cinder cones and other volcanic features located in western New Mexico. The name El Malpais is from the Spanish term *Malpaís*, meaning badlands, due to the extremely barren and dramatic volcanic field that covers much of the monument's area. It was designated a National Monument in 1987 under President Reagan.

El Malpais is part of the Zuni-Bandera volcanic field and is also known as Bandera Lava Field, Grants Malpais, or Malpais volcanic field. It is located on the southeast corner of the Colorado Plateau. The type of lava produced is basalt in many forms: vesicular, aphanitic, pahoehoe, a'a, cinders and scoria.

Zuni-Bandera Volcanic Field containing 15 volcanic vents

Map of the Colorado Plateau showing the El Malpais National Monument on the southeast side

The previous map demonstrates the tectonism of the Flood. Geologists estimate that the Colorado Plateau was uplifted around 10,000 feet. As the Colorado Plateau was raised toward the end of the Flood, it would have produced significant rifting and volcanic activity at its edges. The rifting is evident by the volcanic vents and flows surrounding the Plateau.

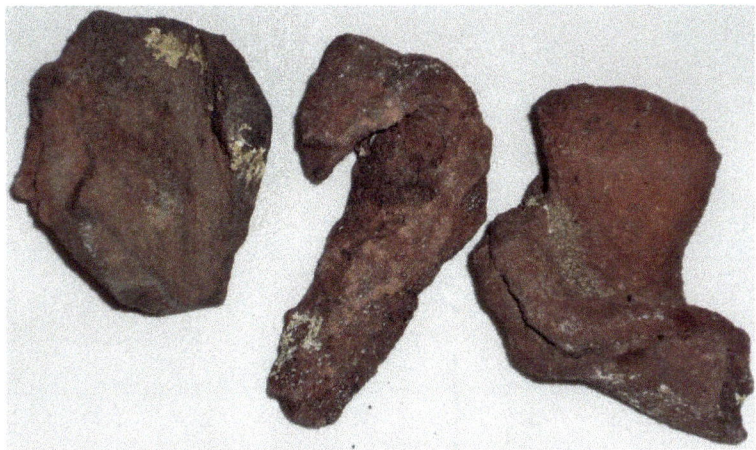

Basalt ribbon bombs, El Malpais

Scoria   Aphanitic basalt with visible olivine crystals

Columnar vesicular basalt showing one of the patters of columnar basalt – five sides: top view.

Pahoehoe lava flow, El Malpais

Oxidized pahoehoe lava, El Malpais

Lava Beds National Monument

Word challenges: lava tube, fumarole, pit crater, hornitos, maars, shield volcano

Lava Beds National Monument is in northeastern California. The Monument lies on the northeastern flank of the Medicine Lake Volcano, and has the largest total area covered by a volcano in the Cascade Range. It became a national monument in 1925 and includes 46,000 acres. Lava Beds National Monument has numerous *lava tube* caves, with twenty-five having marked entrances, and developed trails for public access and exploration.

In addition to the lava tube caves, Lava Beds National Monument includes a variety of other volcanic formations including *fumaroles*, cinder cones, spatter cones, *pit craters*, *hornitos*, *maars*, lava flows, and volcanic fields.

Lava Beds National Monument

Lava Beds National Monument is part of The Volcanic Legacy Scenic Byway. If you want a terrific education in volcanoes and volcanic rocks, this 500-mile drive is a must do! More variety of volcanic formations can be see and studied on this drive than in any other volcanic province in the US. It is one of 31 All American Roads in the US. These are considered the *crown jewels* of the American highway system.

Lava Beds is primarily a collection of basalt flows and cinder cones. Pahoehoe lava is the most abundant lava represented in the monument. Ninety per cent of the lava is basalt; the rest is andesitic. Secular geologists date this province to the outer age of 2,000,000 years old, based on radiometric dating of the lavas. A quick review of our Biblical historical framework, however, will tell us that this age is just not possible. A more precise age would be sometime after the great ice event, within the last 3,500 years. Over 30 separate lava flows make up Lava Beds National Monument. (See Appendix B for more on radiometric dating.)

Lava Beds has the greatest concentration of lava tubes in North America. Some of the lava tubes have speleothems, or crystal formations called dripstone. These formations seem to have formed when (1) lava splashed on the inside walls of the tubes creating lava structures, and (2) minerals leached into the lava tubes from pumice, gravel, soils, and overlying rock by way of rain or snow, depositing the minerals on the lava structures, forming secondary formations, known as secondary speleothems, (dripstone.)

Oregon section of The Volcanic Legacy Scenic Byway

Map of the California section of The Volcanic Legacy Scenic Byway; the best access to Lava Beds National Monument coming from the north is off US Highway 97.

Medicine Lake Volcano (in the background), a **shield volcano**, has produced lava flows that are part of the Lava Beds Monument. It has a 4.3 x 7.5-mile caldera at the center. Shield volcanoes have mild eruptions unlike those of volcanoes like Mount St. Helens.

Thought Questions

1. What is an archipelago? How does it relate to the Hawaiian Islands?

2. Describe the rock basalt. What is its main characteristic?

3. Explain how a lava tube is formed.

4. Describe the difference between a'a lava and pahoehoe lava.

Activity: Go online and search out video presentations of the recent lava eruptions and flows on the big Island of Hawai'i. Write a short report of what you learned.

Lesson Ten – Climate Change and Global Warming
Glacier National Park

Word challenges: glacier, shale, global warming, climate change, stromatolites, cyanobacteria, photosynthesis

Hidden Lake and Bearhat Mountain in Glacier National Park

Glacier National Park

Glacier National Park has been known in the past for its *glaciers* – 150 in the mid-1800s and only 25 in 2010. Something has happened to Glacier National Park! We will discuss this a little later. The park comprises over 1,500 sq. miles with gorgeous, lofty, picturesque mountains, amazing scenic vistas (there are six mountains in Glacier Park over 10,000 feet), and is home to diverse wildlife, including 1,132 plant species, 62 mammal species including grizzlies, wolverine, and wolves, and 260 species of birds. Glacier became a national park in 1910.

What is a glacier? A glacier is a persistent body of dense ice that is constantly moving under its own weight. It forms where the accumulation of snow exceeds its melting over many years, often centuries. Glaciers form only on land and are distinct from the much thinner sea ice and lake ice that form on the surface of bodies of water. At present, glacial ice is the largest reservoir of fresh water on Earth.

The mountains of Glacier National Park are sedimentary rocks from the Belt Supergroup, an almost unbelievable 11-mile-thick water-laid *shale*. This shale gives the mountains of Glacier their unique layered look. The glaciers that formed shortly after the Genesis Flood have further sculpted these mountains into the majestic landforms visible today.

Shale with ripple patterns of the Belt Supergroup, Montana, an indication of a tremendous amount of water

Ripple marks in the shale of the Belt Supergroup

Now for the haunting question of why the glaciers seem to have been disappearing at an alarming rate. Many scientists attribute this rapid melting pattern to *global warming*. Global warming is not just *climate change*. Global warming carries the belief that man-made causes devastate the planet. However, glaciologists point out that the glaciers of Glacier National Park have been melting since the last Ice Age, which ended about 10,000-12,000 years ago, according to secular dating methods. So, in one sense it should not surprise us that the glaciers of Glacier National Park have continued to melt. But these facts are largely ignored because the cause of the Ice Age is not well understood.

If we use our Biblical framework to interpret the formation of these glaciers, then they must have formed after the Genesis Flood about 4,500 years ago. We have already seen from previous lessons that the volcanism and the warmer oceans precipitated by the breaking up of the fountains of the great deep would have been the primary mechanism in bringing about a great ice event shortly after the Flood. As volcanism has subsided and the oceans have cooled significantly since the Flood, it would just make sense that the mighty glaciers which formed as a result would be in a continuous state of decline. Therefore, disappearance of these glaciers should not surprise us. The Earth has continued to settle down over the last 4,500 years.

Glacier National Park

The sedimentary rocks of Glacier National Park indicate formation by a tremendous amount of water activity, as the above picture indicates. Geologists date these rocks at around 800,000 years to 1.6 billion years old, but not through radiometric dating. The main reference point for geologists is the presence of the abundance of fossil *stromatolites* in these sedimentary rocks. Evolutionists have viewed these interesting fossils as primitive and therefore, according to them, would have originated early in our earth's history. They believe that the stromatolites are evidence of *cyanobacteria* and the cyanobacteria gave rise to Earth's oxygenated atmosphere. But this is all interpreted through the uniformitarian framework.

Fossil *stromatolites* occur in significant amounts in Glacier National Park area. Stromatolites are thought to be the remnants of cyanobacteria activity.

Now, there is a conundrum involving cyanobacteria. Why is that? Cyanobacteria are supposed to be some of the most primitive life on our planet, yet they are extremely complex and have supposedly survived for over three billion years.

Cyanobacteria are a phylum of bacteria that obtain their energy through *photosynthesis*. Photosynthesis is the chemical process used in plants to convert energy in the form of sunlight into chemical energy in the form of sugars. The general process involves the combination of carbon dioxide, water and light energy to produce oxygen and carbohydrates. In plants and algae, this process occurs in an organelle called the chloroplast. These processes are extremely complex! Cyanobacteria are not primitive living things and photosynthesis is not a simple process. These are examples of a design that demands and intelligent designer.

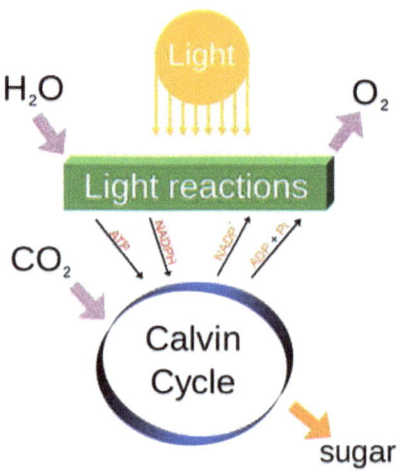

This kind of system speaks of a design that was made beneficial to man from the beginning. Plants were created on Day Three of creation week to prepare the Earth for the existence of other animals, including man. Oxygen is the by-product of this process, and benefits man. In turn, man gives off carbon dioxide that is beneficial to plants.

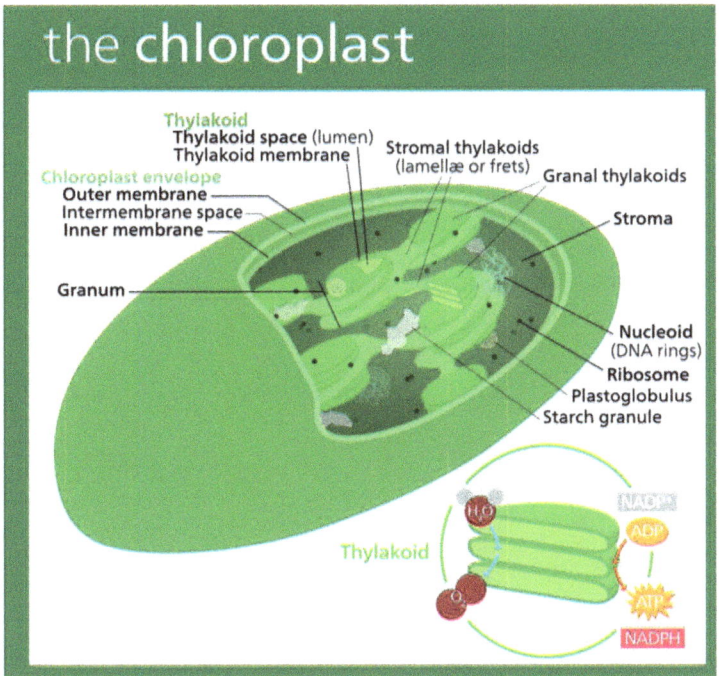

The chloroplast – a complex, well-designed creation specifically designed to do what it does

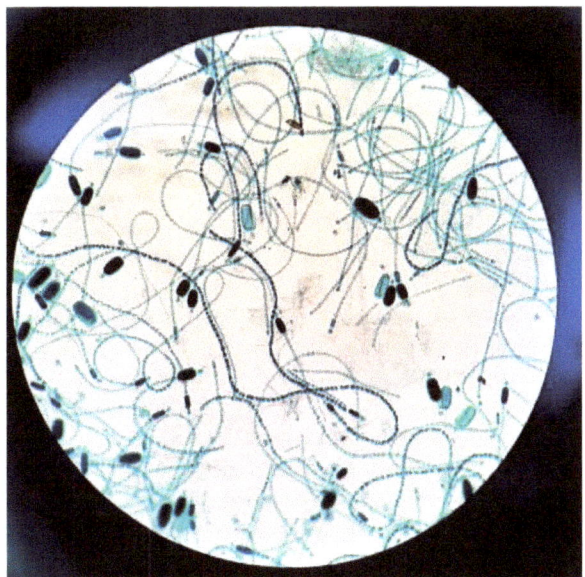

Typical cyanobacteria

Cyanobacteria can be found in almost every terrestrial and aquatic habitat from oceans to fresh water, damp soil, temporarily moistened rocks in deserts, bare rock and soil, and even Antarctic rocks. They are truly ubiquitous and were created to help the environment into which man would be created.

Thought Questions

1. What is necessary for glaciers to form?

2. What is the difference between global warming and climate change?

3. Given the Biblical Framework, where and how do melting glaciers fit in?

4. What is a stromatolite? Why do you think they are so abundant in the Glacier area?

5. What function does the chloroplast fulfill?

Activity. Do a little research on stromatolites. Record *the number of varieties that are in the fossil record. Draw and color a picture of a stromatolite. How do stromatolites fit into a Biblical framework?*

Lesson Eleven – Deposition and Transportation, The Power of Water
Grand Canyon National Park

Word challenges: canyon, geology, science, supernaturalism, igneous, basement rocks, unconformity, metamorphism, hydrological

Grand Canyon National Park
In 1903, President Theodore Roosevelt visited Grand Canyon, and said,

> *The Grand Canyon fills me with awe. It is beyond comparison – beyond description; absolutely unparalleled through-out the wide world... Let this great wonder of nature remain as it now is. Do nothing to mar its grandeur, sublimity and loveliness. You cannot improve on it. But what you can do is to keep it for your children, your children's children, and all who come after you, as the one great sight which every American should see.*

The Grand Canyon was officially designated a national park in 1919.

The jaw-dropping immensity of the Grand Canyon is the first thing that strikes the visitor.

The park covers over 1,900 square miles. As of 2015, the park received more than five and a half million recreational visitors, which is the second highest count of all U.S. national parks after Great Smoky Mountains National Park. Grand Canyon is contained within the State of Arizona. The park is managed by Grand Canyon National Park, the Hualapai

123

Tribal Nation, and the Havasupai Tribe. President Theodore Roosevelt was a major proponent of preservation of the Grand Canyon area and visited it on numerous occasions to hunt and enjoy the scenery. The Grand Canyon is 277 miles long, up to 18 miles wide and attains a depth of over a mile. It is a straight-sided *canyon*, which may give a clue to the rapidity of its formation. We will talk about this later. For thousands of years, Native Americans who built settlements within the canyon and its many caves have continuously inhabited the area. The Pueblo people considered the Grand Canyon (*Ongtupqa* in Hopi language) a holy site and made pilgrimages to it. The first European known to have viewed the Grand Canyon was García López de Cárdenas from Spain, who arrived in 1540.

The Standard Secular Explanation for the Origin of the Grand Canyon
Secularists insist the Grand Canyon was slowly carved over millions of years by the Colorado River. Modern geologists teach that nearly two billion years of Earth's geological history has been exposed as the Colorado River and its tributaries cut their channels through layer after layer of rock while the Colorado Plateau was uplifted. While the specific geologic processes that formed the Grand Canyon and the timing involved is the subject of debate by geologists, they believe that recent evidence suggests that the Colorado River established its course through the canyon at least 17 million years ago. Since that time, the Colorado River continued to erode and form the canyon to its present-day configuration.

The Biblical Framework for Interpreting the Grand Canyon
Let's examine a different perspective. The word *geology* is from two Greek words, *geo* – earth, and *logi* – study of; the study of the Earth. A similar term used for geology is Earth science. Modern geology involves two areas of study in Earth science:

1. The **physical chemistry** – study of the make-up of the rocks, minerals and soils.
2. The **origin** of the Earth – the attempt to explain the history of the Earth in naturalistic terms.

Now, only one of these is *science*. The origin of the Grand Canyon is taught as science. But for something to be labeled as science, it must be observable, testable, and repeatable. So, anything involving the origin of the Earth is not science. It is history and philosophy with some science involved. It is the physical chemistry we can observe, test and repeat. Scientists can melt things, stretch things, examine things, break things apart for observation, but they cannot observe or measure the remote past geological processes of the Earth. They are gone forever. To reconstruct a history of something, written accounts or eyewitness records must serve as the basis for reconstructing it. It is a lot like investigating a crime. And one thing that is crucial to consider when investigating a crime is one's biases. If I am not aware of my preconceived ideas beforehand about a certain person or event, then my conclusions will be skewed. My conclusions can easily be formed around biases that I am not aware of. What's wrong with this?
1. Many scientists are blind to their biases. They are not aware or are not convinced that they even have biases. This will influence how they interpret the physical data and they won't even be aware of it.
2. Many scientists are convinced that their biases are the truth. This is especially troubling when biases are anti-God or anti-Bible. And in the end, it only leads to

blindness and deception for themselves and for others who look to them for answers.

In the teaching of Earth science today, both the physical chemistry and the origin of the Earth are taught as a one package deal. They are both treated equally as science. This is unfortunate and just furthers blindness and stifles creative thinking in young geologists.

So, let's talk about the biases or framework involved in interpreting the history of the Grand Canyon.

A framework consists of the influences I have allowed to form my thinking about things. These influences include religious experience, parental upbringing, friend's opinions, society, my education and believe or not, my personal preferences.

Uniformitarianism

Without a doubt the main bias or framework for interpreting modern geology is the teaching of uniformitarianism. This belief unabashedly declares that the history of the Earth must be framed in terms of what we can observe today happening in nature. No God or global flood was involved and indeed cannot be considered, because of its God-influence, if we are to explain the origin of the Grand Canyon in terms of science, say scientists. The present observable processes have continued throughout the past and they can explain the history of the Earth. So, what is the science in this statement and what is the bias? The science is the observation of the present physical processes, their rates and present effects. What is the bias? It is that no God should be considered in the operations of these processes if we are going to be unbiased and scientific. We must use the same geological processes we observe, to explain the past. Secularists believe that these processes alone are responsible for the rock record of the Grand Canyon. Therefore, for modern geologists, the river flowing down the center of the Canyon, known as the Colorado River, is what produced the huge canyon we see today, because it is what we observe today.

Do you see that we have simply exchanged one bias for another? Even though the Biblical framework and God is historical, it cannot be allowed. That's the bias. Whether I believe God was involved or not is not a matter of science, but of worldview. This is my framework. It is my bias. But we must emphasize here that our biases will have moral consequences. We do not live in a morally neutral universe. This is true for just about everything.

Supernaturalism

The opposite of uniformitarianism is *supernaturalism*. It is the belief that the Hebrew Scriptures record a history of the Earth – its creation and subsequent destruction by a global flood. This belief also involves accepting the witness of the Hebrew Scriptures that a Supreme Being was involved intimately in both the creation of the earth, and the instigation of a global flood that has been responsible for the rock record that we see today in the Grand Canyon.

The conflict between these two frameworks is an ancient story. It had its beginning in Genesis chapter three. And history has repeated it time and time again. It is the story of man's rebellion against his Maker. This story is repeated in the modern geological explanation of the Grand Canyon.

Now that we have briefly looked at the two basic frameworks in interpreting the Grand Canyon, let's apply those frameworks. Yes, it is true that the Hebrew Scriptures can actually explain the origin and nature of the Grand Canyon.

The science in the Grand Canyon would involve the following:
1. The nature of the **rocks** in the Grand Canyon
2. The nature of the **fossils** in the Grand Canyon
3. The present **flow and nature of the Colorado River**
4. The **tectonic forces at work today**

The Nature of the Rocks in the Grand Canyon
What is the Grand Canyon? It is essentially a planed and gouged sedimentary formation consisting predominately of various layers of limestone, shale and sandstone – all of which are types of sedimentary rocks. It also includes metamorphic, plutonic, and volcanic rock. But let's begin our discussion with sedimentary rocks.

The rocks and the incised canyon readily capture an observer's eye. But take another look. The Grand Canyon is essentially a planed surface with a huge gouge. What erosive force would have incised the canyon and yet left a perfectly flat top?

Sedimentary Rocks
Sedimentary rocks are rocks laid down by water and mud. These rocks contain most of the fossils we discover today, from dinosaur bones to clams. And the Grand Canyon is filled with fossils, lots of fossils. Most of the visible Grand Canyon is made up of sedimentary rocks.

Sedimentary rocks are divided into three groups:
(1) The clastic sedimentary rocks, from the Greek word *klastos*, meaning *broken*. These rocks then are made up of broken bits and pieces of rocks and minerals cemented together by a cementing agent like calcite-rich or quartz-rich mud or other sediment.
(2) The chemical sedimentary rocks, made from super-saturated solutions of lime mud, salt or gypsum; and,

(3) Biochemical sedimentary rocks. These rocks have been made from the remains of once-living things like marine creatures and plants.

In the Grand Canyon, these sedimentary rocks include: Sandstone, shale, siltstone – clastic sedimentary rocks; limestone, without fossils – chemical sedimentary rock; and fossil limestone – biochemical sedimentary rock.

Various types of sandstone, all found in the Grand Canyon

Non-fossil limestone   Fossil limestone

Metamorphic Rocks

The word metamorphic comes from a Greek word meaning, *to change form*. These are rocks that have been produced from other rocks through what is thought to have been lots of heat and pressure. No one has ever seen metamorphic rocks form, so we must guess as to their origin, using our biases, framework, or worldview. The primary metamorphic rock in the Grand Canyon is called the Vishnu schist. Schist is a mica-rich rock. The Vishnu schist, in combination with plutonic rocks, form what geologists call, the basement rocks.

Schist

Plutonic Rocks

Plutonic rocks include the granites. There is a granite formation found at the bottom of the layers in the Grand Canyon, called the Zoroaster Granite. Granite is called an *igneous* rock by modern geologists because they *think* it formed from hot liquid magma millions of years ago and then slowly cooled to produce the larger observable crystals, typical of plutonic rocks. But do geologists know this for sure? No one has ever seen a granite in the process of forming! That may surprise you. Well, if geologists have never seen granite in the process of forming, then how do they know it is an igneous rock? They don't. It is a guess, based on their uniformitarian framework, that has been taught as fact. Therefore, some like to call rocks like granite, plutonic rocks, after the mythical god of the underworld, Pluto. Plutonic rocks are considered by all geologists to be the basement rocks.

Granite

Volcanic Rocks

According to geologists, over 150 flows of basaltic lava dammed the Colorado River at least 13 times throughout the Canyon's formation. Deciphering differences in lava flows is a matter of *chemistry*. Deciphering the history of volcanic lava flows is a matter of *worldview*, unless of course these flows have been observed, as in the case of some on the Big Island of Hawai'i. What is interesting about the Grand Canyon flows, however, is their radiometric dates. Dating of the Grand Canyon basalt has been all over the geologic dating map. (These dates and dating methods can be researched by going to Creation.com and reading some of the articles by typing the key words, "dating of lavas at Grand Canyon" in the search box.)

The word basalt means *hard*. It is a very hard volcanic rock. It is dark in color because of the darker minerals that form it. It is typically a lava that is very hot, erupting generally in mild showers or in flows at temperatures as high as 2,000°F. Basalt is very low in quartz, generally around 20% of its make-up. This makes the basalt lava non-viscous, which means that it has very little resistance to flowing. It can flow very quickly and over large areas of surface before cooling and ceasing in its activity.

What is the origin of basalt lava? Uniformitarian geologists teach that basalt helped to form the crust of the planet Earth around 4.6 billion years ago and has been forming and reforming the Earth ever since. In fact, a recent National Geographic article is entitled, "Volcanoes: Crucibles of Creation." Modern geologists teach that the early Earth was a planet of volcanism. This framework helps modern geologists explain the presence of so much lava spread around the Earth. Two volcanic provinces that contain this type of lava are the Deccan Traps (multiple layers of lava over 6,000 feet thick, and 193,051 square miles) in India, and the Columbia Plateau in Washington, USA (lava that is over 6,000 feet thick and covering an area of some 63,000 square miles). There are only two ways to explain this kind of outpouring of lava. Either it was erupted rapidly over a short period of time or it erupted slowly over a long period of time. It is only by radiometrically dating these lavas that old ages for these lavas have been obtained.

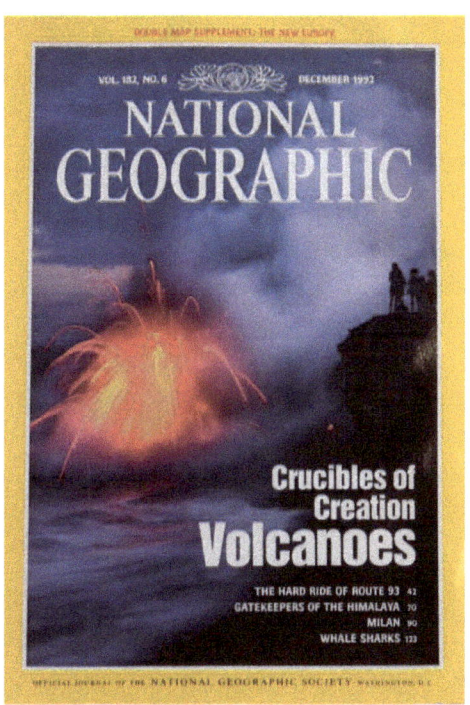

The Grand Canyon area is filled with volcanic remnants, some of which are still very active. Arizona has documented over 600 volcanoes across its state. One of the most famous areas is the San Francisco Volcanic Field, just southeast of the Grand Canyon.

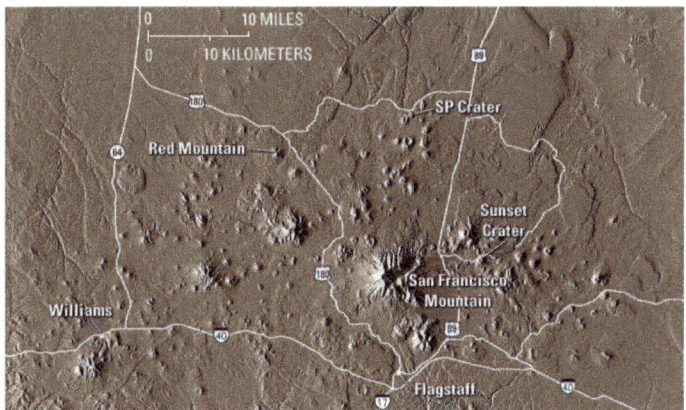

The San Francisco Volcanic Field

Basalt: a dark volcanic rock low in quartz and high in iron

The Nature of the Fossils in the Grand Canyon

Fossils are preserved remains or traces of plant and animal life. These fossils tell us what kinds of plants and animals were buried in sedimentary rock. They do *not* tell us how old these plants and animals are. The interpretive framework secular geologists use to determine their age is evolution. Biblicists think because all the fossils found in the Grand Canyon are contained within sedimentary rocks, a plausible interpretation is that these sediments with billions of fossilized remains were all laid down during the year-long Genesis Flood. Some of the fossils recorded in the Grand Canyon include the following (representative examples):

 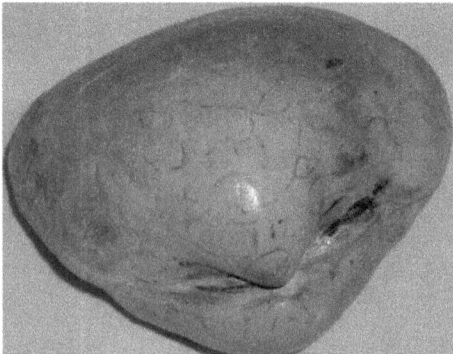 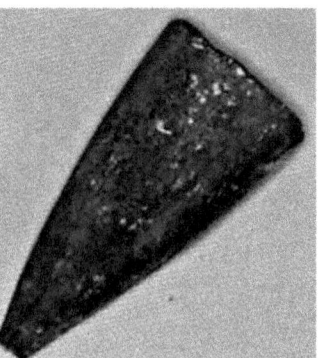

Fossil bryozoans Fossil clams Fossil squids

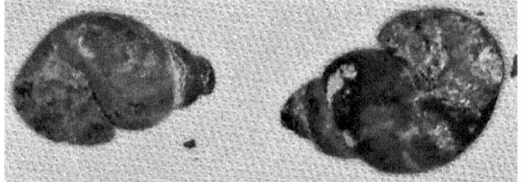

Fossil snails, called gastropods

Fossil snail, called a gastropod

Trilobites   Fossil corals   Fossil brachiopods

Fossil crinoids   Worm borrows

Tracks, called ichno fossils, leaves and ferns

One of the greatest mysteries of the Grand Canyon is the presence of stark *unconformities*. The Grand Canyon is filled with unconformities. What are these? A general geology dictionary definition reads like this:

> An **unconformity** *is a buried erosional or non-depositional surface separating two rock masses or strata of different ages, indicating that sediment deposition was not continuous. In general, the older layer was exposed to erosion for an interval of time before deposition of the younger, but the term is used to describe any break in the sedimentary geologic record.*

Pictured below is a simple cut-away sketch of the Grand Canyon showing the various layers of rock names and the geological Periods in which they were supposed to have formed.

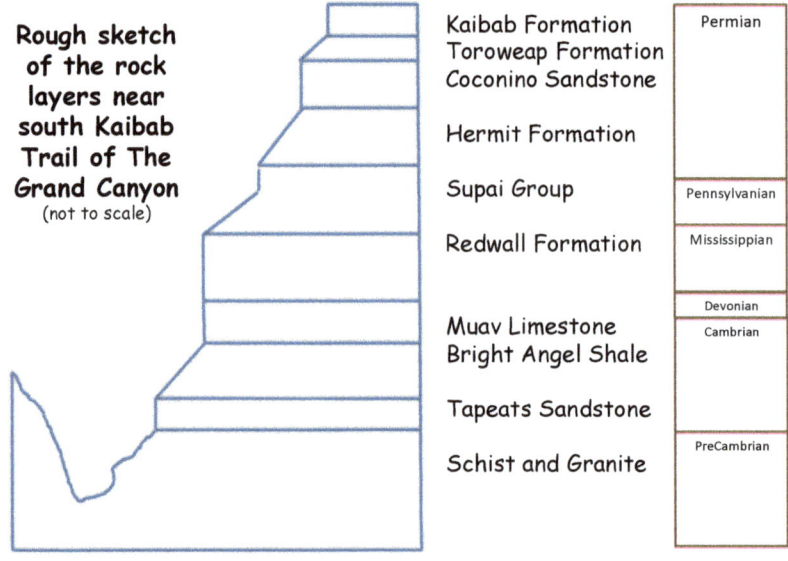

The layers look continuously laid down and the contacts between the layers are mostly flat contacts. But geologists tell us that there are massive periods of time/events missing between the layers. The reason for this is because of the belief in the evolution of all living things. They believe that the rock layers in the Grand Canyon are supposed to represent stages in evolution laid down over vast eons of time – except that some of these stages are missing. The fossils that are supposed to be present are not, and they know this. Secular geologists tell us that there are nine unconformities representing about 1.3 billion years of missing time. How can one construct a history missing so much of the story? Would it not be more representative to reconsider the record of Earth history as recorded in the Bible using the Flood framework? Doing so would give a good explanation for the way the layers are laid down – in successive order and with flat contacts between the layers – exactly the way the look in the field! A composite view of the Grand Canyon shows us basement rocks, plutonic rocks, that originally formed the crust of the Earth, then were uplifted during and shortly after the Flood, causing *metamorphism* and then sediments were layered on containing billions of dead things, laid down successively and rapidly in wave after wave of the flooding of Genesis mentioned in chapters seven and eight.

In this scenario the Colorado River would not have been the cutting agent that formed the big gash in the sediments, but the remnant of the water involved in the channelized stage of the retreating floodwaters. (See Lesson Seven.) I mentioned earlier that the Grand Canyon has straight-sided walls. Straight-sided walls are typically caused by rapidly moving water cutting through sediments in a short period of time. This has been observed in modern floods. The cutting action at Grand Canyon was caused by water moving rapidly through soft sediments. It is a very simple explanation, and it also fits our Biblical framework. This was not accomplished by a gradual river flow over millions of years.

This picture shows that a lot more water would have been needed to cut these sediments, which would have been freshly laid during the flood stage of the Genesis Flood. A river would not have cut this.

The Bible speaks of a global, devastating flood that would have carved up the landscape, transporting and depositing layer after layer of sedimentary rocks. The Flood was a powerful *hydrological* event of unimaginable proportions. It alone can explain the catastrophism that is evidenced in the Grand Canyon.

The Grandness of Grand Canyon: Facts of Interest

- Grand Canyon established as a forest preserve in 1892
- Became a national monument in 1908
- Designated a national park in 1919
- National park enlarged in 1975
- Total acreage: 1,215,375
- Length in air miles: 190
- Length in river miles: 277
- Minimum width: 600 feet at Marble Canyon
- Maximum width: 18 miles
- Average width: 10 miles
- Minimum width of Colorado River: 76 feet
- Average width of Colorado River: 300 feet
- Maximum depth of river: 85 feet
- Average depth of river: 40 feet
- Number of rapids: 160
- River average gradient: eight feet per mile
- Total miles of trails in the park: 400
- Total number of those trail miles that are maintained: 30.7
- Number of miles of roads: 355
- Number of bird species in the park: 287
- Number of mammal species: 88
- Number of reptile and amphibians species: 58
- Number of plant species: 1,500
- Number of fish species: 26
- Number of biotic life zones in the park: five (the same as traveling from Mexico to Canada)
- Endangered wildlife species: three (bald eagle, peregrine falcon, humpback chub)
- Endangered plant species: two (Brady pincushion cactus and sentry milkvetch)
- Number of known archaeological sites within the park: 2,700
- Number of Indian reservations in the park: One (Havasupai)
- Number of buildings listed as National Landmarks: 120
- Number of structures on National Register of Historic Places: 136

Thought Questions

1. How big is the Grand Canyon?

2. What is it about the canyon that indicates formation by a watery catastrophe?

3. Define the word *geology*. What do secularists include in the definition of the word?

4. Define the word *science*. What should not be included in the definition of science?

5. What is an unconformity? What geological proof is there for unconformities? What philosophical proof is there for the existence of unconformities?

6. What is hydrology? How was the Genesis Flood a hydrological event?

Activity: Obtain a gallon glass jar or another large glass container. Fill it with a mixture of various sized pebbles, grains and soils. Fill it with water and shake it up. Observe what happens until all the sediments have settled. Write your observations down.

Lesson Twelve – The Rock Types, How Much Have We Observed?
North Cascades National Park

Word challenges: massif, breccia, diorite, granodiorite, gabbro, schist, quartzite, gneiss, serpentinite, clast

North Cascades National Park

North Cascades National Park is a complex of two units within the one national park. It is located in north-central Washington, USA. North Cascades National Park is the largest unit of the park complex. It is divided into a North Unit and a South Unit, bisected by the Ross Lake National Recreation Area. The park features numerous mountain peaks and glaciers.

The Cascade Range of Mountains have been called the most complex mountain system in the world. This is because there are so many different features of the mountains – from volcanoes, to metamorphic *massifs,* to plutonic provinces of granites, to sedimentary rocks, spread over the Cascades. Every rock type is represented in the Cascades and it is difficult to sort it all out. Many geologists have tried.

North Cascade Range is considered the most rugged mountain range in the contiguous United States.

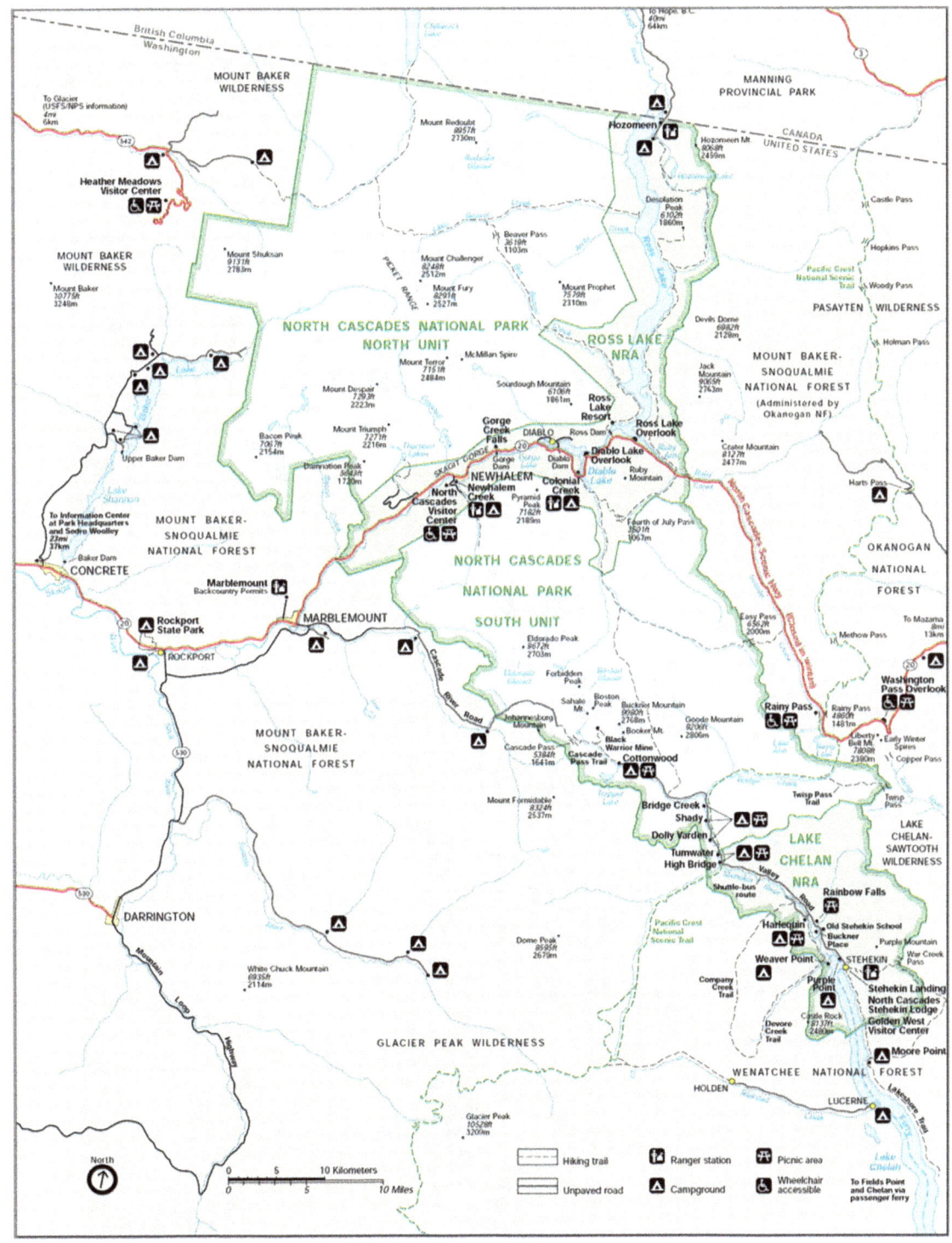

Map of the North Cascades National Park

North Cascades National Park was incorporated as a national park in 1968. It has about 21,000 visitors each year, mostly hikers. The terrain is rugged, and the peaks are jagged, due to past glacial activity.

Pelton Peak

The Pickett Range is the most rugged subrange within the North Cascades.

Glacier Peak, a stratovolcano, is the dominant feature in the southern portion of the North Cascades National Park.

Mt. Baker is one of the loftiest and most conspicuous stratovolcano peaks just west of North Cascades National Park.

Some of the present peaks of the Cascades seem to have been built on top of mountains that most likely erupted late in the Flood Period or shortly after. Glacial activity has eroded much of the former peaks, so the newer volcanoes are less than 4,000 years old. The present-day Cascade Volcanoes are considered young geologically.

Sampling fumarole gas at Sherman Crater on the south side of Mt. Baker in 1981

Mount Baker has the second-most thermally active crater in the Cascade Range after Mount Saint Helens.

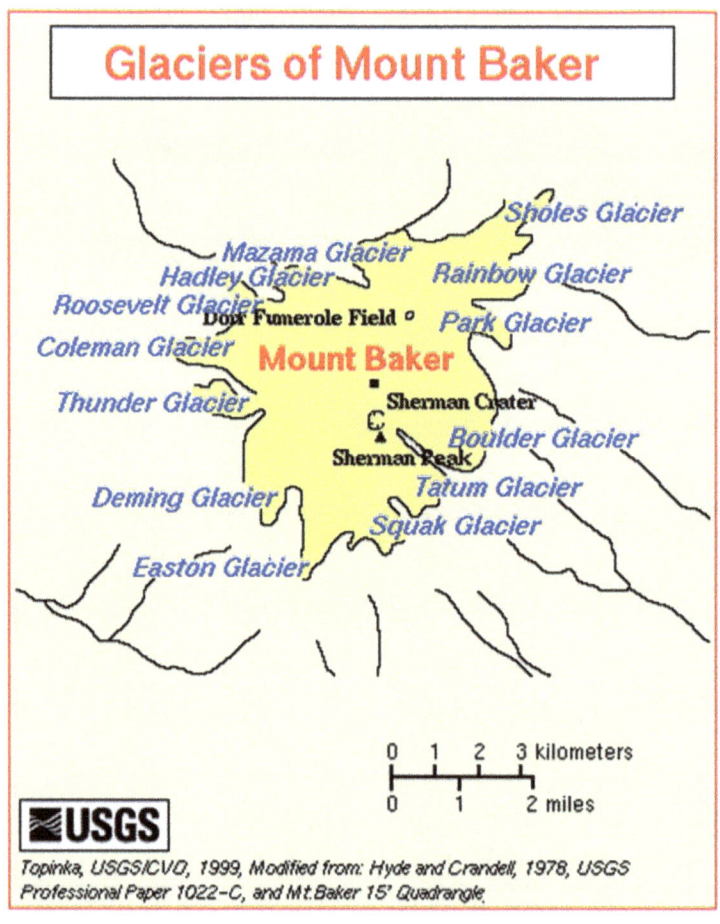
Map of Mt. Baker, identifying 13 glaciers

Mt. Baker is the most heavily glaciated peak in the Cascade Range. The glaciers of Mt. Baker have shown an interesting pattern over the years. All retreated during the first half of the century, advanced from 1950–1975 and have been retreating increasingly rapidly since 1980. This might have something to do with the increased thermal activity in the volcano. It is one of the most studied of the Cascade volcanoes and is considered an active volcano.

The Rocks of North Cascades National Park
The volcanic rocks consist of volcanic *breccia*, andesite, basalt, and rhyolite. Much like the sedimentary conglomerate, a volcanic breccia consists of angular volcanic rocks (basalt) cemented within a volcanic mud flow. The most ubiquitous rock of the stratovolcanoes is andesite, an intermediate volcanic rock usually with an abundance of sodium feldspar in its groundmass.

Volcanic breccia

Andesite – typical of the Cascade volcanoes   Rhyolite – not common in the Cascades

Basalt in columnar form: this is not common in the Cascades. This is part of the Columbia Flood Basalts found on the plains to the immediate east of the Cascades.

Another prominent mountain in the North Cascade Range is Mt. Shuksan, which is a glaciated *massif*. The word is taken from French, (meaning massive) where it is used to refer to a large mountain mass or compact group of connected mountains forming an independent portion of a range. Mt. Shuksan is thought to be a section of the Earth's crust that was pushed up and then glaciated. And although geologists date this mountain over 150 million years old because of radiometric dating, fitting it within the Biblical framework would place this up-thrusted mountain at the time of the Genesis Flood, when the fountains of the great deep burst open. It is composed of plutonic basement rocks and metamorphic rock. Later it was glaciated and sculpted by the glacial event immediately following the Flood.

Mt. Shuksan, just east of Mt. Baker, is not a volcano.

Plutonic rocks are those in which the mineral grains can be easily seen with the naked eye. The three most prominent kinds of plutonic rocks in the Cascades are these:

Diorite   Gabbro   Granodiorite

Diorite is an intermediate plutonic rock consisting of an even mix of plagioclase feldspar (the white feldspar) and the dark rock-forming minerals biotite and hornblende. *Gabbro* contains some of the same minerals, but the feldspar is calcium. *Granodiorite* is much like diorite, except that it contains quartz.

Metamorphic rocks exhibit change. *Schist* contains a lot of mica. Its parent rock is thought to be sandstone. *Quartzite* is thought to be metamorphosed sandstone. In most quartzites one can still see the layering from the parent sandstone, as in the picture below. *Gneiss* consists of banded dark and light minerals and thought to be metamorphosed granite. *Serpentinite* is a dark green rock and its origin and parent rock is not clear. It does contain the mineral serpentine, however. All four of these rocks are abundant in the Cascades.

Schist   Quartzite

Gneiss   Serpentinite

How were these rocks formed? How old are they? The scientific answer is, we don't know. Since no one has seen them form, except volanic rocks under present circumstances, there is no scientific answer to the question. The answer lies in proposing a history or philosophy of rock formation using an interpretive framework – either uniformitarianism or Biblical creation/catastrophism. Secular geologists use an illustration to help organize the rocks into an explaination about their formation. This device is called The Rock Cycle. It could be quite instructive – if it were true. But in studying it the question naturally arises. How much of it has been observed?

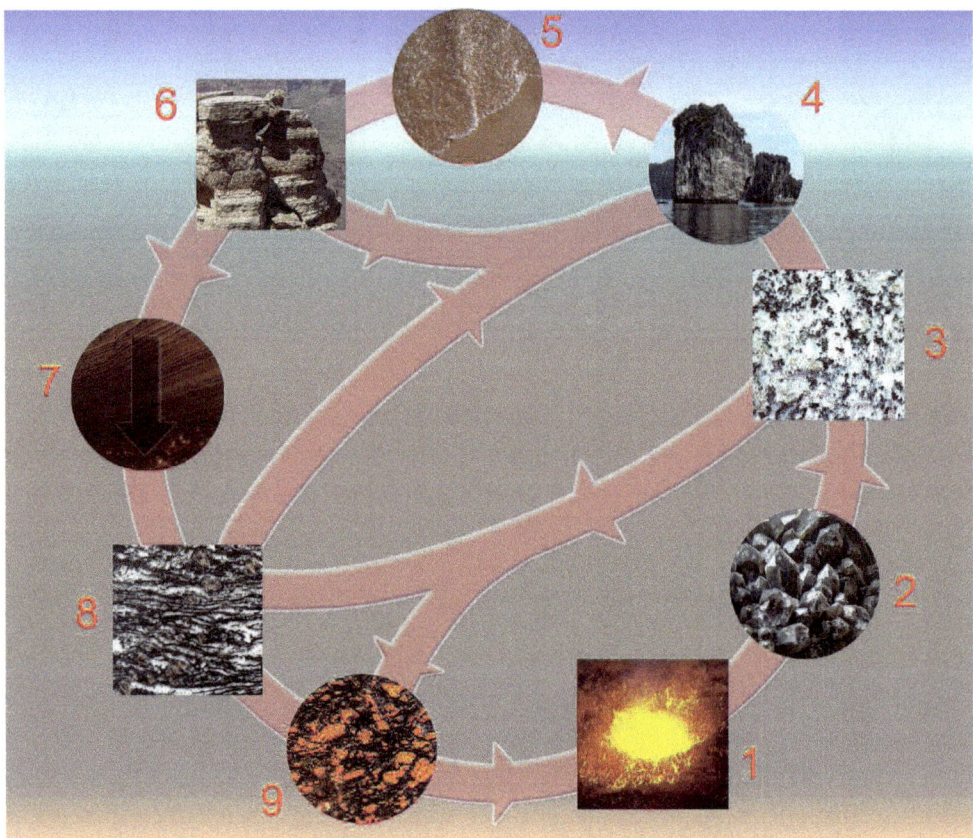

The Rock Cycle as envisioned by uniformitarians

The illustration of the Rock Cycle is easy to understand. Secular geologists believe that Earth's early rocks all came from magma. The magma crystalizes into rocks like granite. These are grouped as igneous rocks because secular geologists believe that all rocks like granite were once molten magma and that through millions of years of cooling, formed the igneous rocks. In some cases over millions of years, igneous rocks can deform into metamorphic rocks. But the igneous rocks can also break down through slow and gradual erosion to become sedimentary rocks. They then are eroded further and through either metamorphism, melting, or sedimentary processes become other kinds of rocks. While this might be theoretically true, not much of it has actually been observed. But with the rejection of the Flood and its geology, there isn't anything to take the place of The Rock Cycle.

So, let's take a different approach. Secular geologists organized the rocks base on how they *believe* they were formed – their uniformitarian history. Let's organize the rocks on the basis of their chemistry and physical properties. How would the rock types now be organized?

We will first separate the igneous rocks: the first is those formed by heat and fire, into **volcanic rocks**. We have witnessed the formation of many of the volcanic rocks. So those are the true igneous rocks because of their formation by fire. Another defining charactristic of volcanic rocks is that they are *fine-grained*. That is, one cannot easily see the mineral crystals that form them. We can generally tell what kind of volcanic rock they are because of the color of the rock. The color is derived from the minerals that form them.

The other so-called igneous rocks, **the plutonic rocks**, will be organized based on the mineral grains or crystals that can be easily seen with the naked eye. We won't call them igneous, because we haven't seen them form from fire. These rocks are also called, *coarse-grained* rocks, because their minerals can easily be seen. This is not a commentary on how they were formed, but their observable characteristics.

The sedimentary rocks will be organized based on the varying sizes of *clasts* (broken bits and pieces of other rocks or minerals) that form them. They can also be organized according to the biomaterials that form many of them, like coal.

The metamorphic rocks will be organized based on whether there is banding or layering, like gneiss, or crystalization, like quartzite or marble.

So, we will end up with four types of rocks instead of the three that secular geology recognizes.

Do the rocks tell us anything about the origin of the Earth or how old it is? Aside from radiometric dating, there is no way to propose an age of the Earth or to explain its history by looking at the rocks or fossils in them. And radiometric dating relies on several assumptions about the way radioactivity has worked in the remote past. It is not a reliable system to use for the determination of the age of the rocks or Earth. Any age based on fossils relies on the belief that life evolved and therefore there are primitive fossils on the bottom of the rock heap and more complex fossils on the top of the rock heap. You can read more about this in Appendix B.

Thought Questions

1. What accounts for the jagged appearances of the mountains that form the Cascade Range of mountains?

2. Why is the Cascade Range called a complex range of mountains?

3. Why should we not use The Rock Cycle? How would a Biblical framework organize the rock types?

4. What are the distinguishing characteristics of volcanic rocks? Of plutonic rocks? Of sedimentary rocks? Of metamorphic rocks?

5. What is a massif? Give an example of one in the Cascade Range.

Activity: Draw an illustration of a geological scene (for instance, include such features as rifts, volcanic eruptions, mass burial of animals, planation, channelization, etc.) that incorporates as many features of the Flood framework as possible. On another piece of paper, explain the features you used and why.

Lesson Thirteen – Interpretive Frameworks and Biases:
Mount St. Helens National Volcanic Monument

Word challenges: stratovolcano, lahar, pyroclastic, ash, dacite, catastrophic, analog

Mount St. Helens stratovolcano, looking at the north side of the volcano

Mount St. Helens

Mount St. Helens is a ***stratovolcano*** with an elevation of 8,363 feet. It is located in southwest Washington, 96 miles south of Seattle and 50 miles northeast of Portland, Oregon. Mount St. Helens was named for the British diplomat Baron St. Helens, a friend of explorer George Vancouver who made a survey of the area in the late 18th century. The volcano is part of the Cascade Range of mountains that run 700 miles north to south, beginning in southern British Columbia, Canada and ending in northern California at Mt. Lassen. All of the eruptions in the contiguous United States over the last 200 years have been from Cascade volcanoes. The two most recent were Lassen Peak from 1914 to 1921 and a major eruption of Mount St. Helens in 1980. Minor eruptions of Mount St. Helens have also occurred since, most recently from 2004–2008 with continuing volcanic activity into the present.

360° panorama photo of Mount St. Helens crater rim from the south in 2009

Map of the Cascade Range shown in the states of Washington, Oregon and California and in British Columbia, Canada; Mount St. Helens lies directly west of Mt. Adams, another stratovolcano.

The Mount St. Helens 1980 eruption has gone down as the deadliest and most economically destructive volcanic event in the history of the United States. Fifty-seven people were killed; 250 homes, 47 bridges, 15 miles of railways, and 185 miles of highway were destroyed. A massive debris avalanche, known as a *lahar*, triggered by an earthquake measuring 5.1 on the Richter scale caused the 1980 eruption that reduced the elevation of the mountain's summit from 9,677 feet to 8,363 feet, leaving a one-mile wide horseshoe-shaped crater. The lahar was up to 0.7 cubic miles in volume. The Mount St. Helens National Volcanic Monument was created after the 1980 eruption to preserve the volcano and allow for its aftermath to be scientifically studied.

(Left) Mount St. Helens before the 1980 eruption - it had earned the nickname, *The Mt. Fuji of America*.
(Right) This picture was taken four months after the eruption from the same location as the top photo.

The massive *blow-down* of the forests on the volcano

The massive *pyroclastic* mud/ash flows stripped the trees and laid them down like a box of toothpicks.
Notice the height of the mud/ash flow preserved on the trees in the picture.

An interesting phenomenon appeared in the winter of 1980-1981: a new glacier appeared, now called Crater Glacier.

Crater Glacier visible just above the new dome within the Mount St. Helens crater formed in 1980

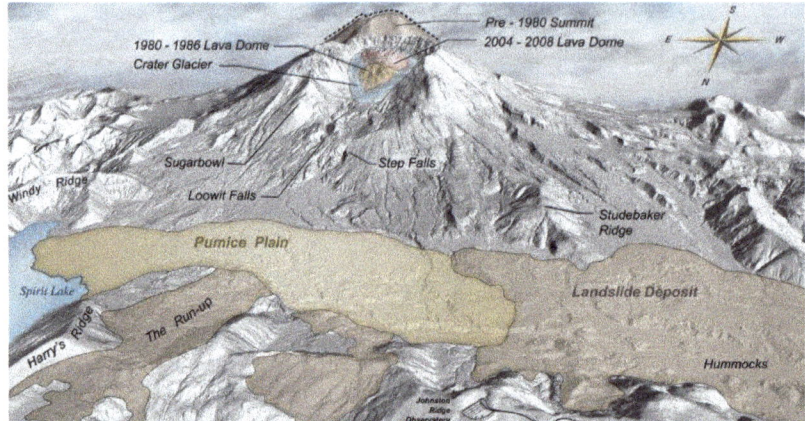
Drawing of prominent points of geological interest at Mount St. Helens

Two views of Mount St. Helens from the same point nine miles north of the volcano - 1979 and 1980

Typical for stratovolcanoes, the main rock of Mount St. Helens is andesite. Andesite draws its name from the type mountains of the Andes in South America that are made of the same volcanic rock. A type mountain is usually one that was first studied, incorporating a typical rock, in this case, andesite. Andesite is an intermediate colored rock between *dacite* and basalt – light to dark gray in color. It contains more quartz and sodium feldspar than basalt does, but less quartz than dacite.

Andesite   Dacite

Ash from Mount St. Helens; the name *ash* is a bit misleading. It is not like the ash from burnt paper or wood. Volcanic ash consists of tiny rock particles and glass. In a pyroclastic flow it acts like a sand blaster and easily removes bark and limbs from trees.

Over 2.4 million cubic yards of *ash*, equivalent to about 900,000 tons in weight, were removed from highways and airports in Washington State. About 250,000 cubic yards of ash were stockpiled throughout the state for future use. Interstate 90 from Seattle to Spokane, Washington, was closed for a week, following the eruption of 1980. Air transportation was disrupted for a few days to two weeks as several airports in eastern Washington shut down due to ash accumulation and attendant poor visibility. Over a thousand commercial flights were canceled.

The fine-grained, gritty ash caused substantial problems for internal-combustion engines and other mechanical and electrical equipment. The ash contaminated oil systems, clogged air filters, and scratched surfaces. Fine ash caused short circuits in electrical transformers, which in turn caused power blackouts. The sewage-disposal systems of several towns that received about half an inch or more of ash, such as Moses Lake and Yakima, Washington, were plagued by ash clogging and damage to pumps, filters, and other equipment.

Ash fall from Mount St. Helens, 1980, at noon in Yakima, Washington just east of Mount St. Helens

The Mount St. Helens eruption was horribly destructive. Scientists thought it would take many years for life to once again thrive on the devastated and burned out landscape. In fact, many thought it would never happen. To everyone's amazement, plant and animal life started to return very soon after the eruption. It is a testament to God's wonderful creation and the resilience that He gave plant and animal life.

Plant life began appearing very early on after the eruption. This picture was taken in 1984.

Mount St. Helens hillside in 2005, 25 years after the eruption. Today, the logs are all but gone, and vegetation is thriving.

A striking observation of Mount St. Helens was the vast number of trees that were stripped and buried as a result of the explosion. A similar observation was made concerning the trees that make up an area called Specimen Ridge in Yellowstone. An abundance of fossil trees was discovered there in the late 1800's, in all kinds of positions, buried in layers of volcanic ash, soil, and rock. The uniformitarian framework interpreted these trees as *successive forests* that grew in place there over many thousands of years. At the time this interpretation was used successfully to refute the Biblical claim that the Earth was just 6,000 years old.

This uniformitarian framework was applied early in the interpretation of Specimen Ridge at Yellowstone National Park. Specimen Ridge was referred to as *Specimen Mountain* by local miners and was probably named by prospectors well before 1870. Specimen Ridge is located north of Lamar Valley in northeastern Yellowstone Park.

Lamar Valley below the hills that contain Specimen Ridge

Specimen Ridge 1890 and a copy of the original drawing proposed by an early geologist of Yellowstone in the late 1800s

Tree at Specimen Ridge today

This interpretation held sway in geological schools until the 1980s when the eruption at Mount St. Helens took place in western Washington, which demonstrated that trees could have rapidly been torn up and reburied in layers of volcanic ash and mud flows in what would later look like successive forests. In reality, the whole mess was deposited rapidly by the eruption of Mount St. Helens and its subsequent mudflows. The interpretation of Specimen Ridge was subsequently changed in most modern geology books and in Yellowstone Park itself at the interpretive signs for Specimen Ridge. But the damage stemming from the uniformitarian framework had its effect on millions of people for over a hundred years!

The log mat that was washed into Spirit Lake by the Mount St. Helens eruption and mudflows: the volcano is visible in the background.

The Toutle River Badlands

Look at the picture above. This photo is of The Toutle River Badlands. "Where is that?" you ask. You might even think that this picture is of some ancient landscape carved out long ages ago. But it wasn't. This picture was taken shortly after the Mount St. Helens eruption of 1980. This landscape was carved out by the lahars that proceeded from the eruption. This landscape was formed rapidly and by a huge volume of water, mud and rock. Again, a local catastrophe can serve as an *analog* to illustrate the rapid destructive power of the Genesis Flood. An analog is a likeness of a bigger idea or event.

The Toutle River Badlands with Mount St. Helens in the background

The Toutle River Badlands

Geologists have changed their cherished uniformitarian interpretations since the 1800s primarily because of local events like this, that was *catastrophic* in nature. This kind of catastrophism is precisely what the book of Genesis teaches in Genesis chapters 7-8. The flood as recorded in Genesis was catastrophic – globally! And although the Biblical Flood was global, its geological implications are perfectly in line with effects caused by local and destructive catastrophes of today. Mount St. Helens serves as a modern analog of global catastrophism and a small picture of what the Flood must have produced.

Thought Questions

1. What kind of a volcano is Mount St. Helens?

2. How does Mount St. Helens compare to Yellowstone?

3. What determined the interpretation of the fossil trees at Specimen Ridge in the late 1800s? How would it have served to undermine the Biblical reference to a 6,000-year-old Earth?

4. What is an analog? In what ways does Mount St. Helens serve as an analog of the Genesis Flood?

5. What was the main surprise about the aftermath of the Mount St. Helens eruption of 1980?

Activity: Look at some film footage of the eruption of Mount St. Helens. What kinds of things happened that might tell you similar things that the global Flood of Genesis could have accomplished?

Lesson Fourteen – Dinosaurs and The Flood
Dinosaur National Monument

Word challenges: quarry, disarticulation, death pose, bone beds, graveyard fossils, extinction

Dinosaur National Monument

Dinosaur National Monument is a U.S. National Monument located on the southeast flank of the Uinta Mountains on the border between Colorado and Utah at the confluence of the Green and Yampa Rivers. Most of the monument area is in Colorado, with a small portion in Utah. The area was declared a National Monument in 1915.

The rock layer enclosing the fossils is limestone, sandstone, and conglomerate, known as the Morrison Formation. Secular Geologists date this from the Jurassic Period some 150 million years old, because of their belief in evolution. At the same time geologists also acknowledge that these sandstones and conglomerates have a watery origin, generally attributed to a localized flooding event. But because of the Biblical historical framework, I would date the Morrison Formation as from the one global flood period of 4,500 years ago and the "localized" flooding event, also to the Genesis global Flood.

Green River Canyon in the Dinosaur National Monument, part of the Colorado Plateau

Because of the nature of the 800 fossil sites (*bone beds*) and the amount of *disarticulation*, as well as the uniformity of formation, a rapid, catastrophic burial is a better explanation, and preferred to *small and slow and long ago*.

Turtles, crocodiles, sauropods, theropods, ornithopods, stegosaurs and fish have been found in the monument. Conspicuously missing, however, is fossil evidence for the abundance of plant life needed to support this array of creatures. Instead of viewing these finds as representative of the environment that used to live here millions of years ago, it would be better to see these creatures as having been washed in to the area from some other area by raging floodwaters and then buried in a disarticulated state.

Dinosaur *quarry* in the Dinosaur National Monument: notice the amount of disarticulation of the bones in the sedimentary rock.

The **death pose** (*opisthotonus* or *opisthotonos*, from Greek roots, *opisthen* meaning *behind* and *tonos* meaning *tension*,) is a state of severe hyperextension and spasticity in which a creature's head, neck and spinal column enter a complete bridging or arching position. This abnormal posturing is supposedly caused by spasm of the axial muscles along the spinal column. No one is 100% sure of what caused this in the dinosaur and bird fossils we find in abundance in the fossil record. However, it appears that the victims suffered a violent death and the Biblical Flood would certainly explain why. As creatures were trapped alive in raging waters and sediments, many were caught in their final throes of death and preserved as fossils.

Although complete skeletons are extremely rare, necks and heads arched back into what is called, *the death pose*, are not uncommon.

The Morrison Formation encompasses an area of a staggering 700,000 square miles from New Mexico into the provinces of Canada. This formation includes layers of a silica-rich volcanic ash that once covered the whole area. The remains of dinosaurs, clams, and snails are found jumbled within a pebbly (conglomerate) sandstone layer. The Morrison Formation is one of the best examples of catastrophism in geology. It is the perfect picture of billions of dead things buried in sediments and preserved as fossils since the great Genesis Flood.

A fossilized leg bone buried in sedimentary rock

Dinosaur National Monument, *The Wall of Bones*, as it is called, illustrates the effects of catastrophic death and burial.

The term used to describe these collections of disarticulated bones in sedimentary rock is *bone beds* or *graveyard fossils*. (See Lesson Three about disarticulation.) These are common in the fossil record. They are not evidence of individual natural death and slow decay until burial. These bone beds tell us of a watery, muddy, catastrophe. These bone beds demonstrate a consistency of pattern left all around the world. The only satisfactory explanation is that these dinosaurs died, were buried rapidly by sediments, torn up by more raging floodwaters and transported, perhaps great distances. Their disarticulated state tells us that they did not die naturally and then were buried gradually. To avoid decay and scavenging, even bones must be buried quickly. The fact that we have so many of these bone beds and disarticulated fossils, implies a universal catastrophic watery end to these creatures.

Extinction

Why did the dinosaurs die out? What could have happened to bring about what appears to be a sudden disappearance after secular paleontologists tell us that the dinosaurs were some of the most successful, evolutionary creatures, having lasted for about 200 million years? This has been one of the greatest mysteries of modern geology. Many reasons have been proposed through the years, but not one of them has been totally accepted by scientists.

What is *extinction*? The dictionary defines extinction as the end of an organism or of a group of organisms, normally a species. The moment of extinction is generally considered to be the death of the last individual of the species. Secular geologists place this end for the dinosaurs at 65 million years ago. Before 1980 that end had supposedly come about gradually due to perhaps a combination of reasons, but never one of catastrophic proportion. Ever since Charles Lyell in the early 1800s, geologists were contemptuous of the word catastrophism and would not use the word in their literature. It smacked too much of Noah's Flood.

Normally extinction is viewed as a gradual process due to natural or man-induced causes. For many years, environmental and climate changes were the primary catalyst for the decline and extinction of the dinosaurs. It was taught that these were also the main reasons for the rise of the mammals.

Geologists use a device called the Geological Time Table to illustrate the history of the Earth and life on it. The Table is divided into eons, eras, periods, and systems or epochs. The Era that includes the Age of Dinosaurs is the Mesozoic Era, The Era that includes the Age of Mammals is the Cenozoic Era. Mesozoic means *middle life* and Cenozoic means *recent life*. The line between the Mesozoic and the Cenozoic is sharp and clearly defined. It is at precisely 65 million years ago that the dinosaurs were supposed to have vanished and the mammals began to thrive. This is the mystery! How could this happen? Because of their views on how Earth history and the evolution of life were supposed to have occurred, secular geologists could not fathom how the sudden disappearance of the dinosaurs and subsequent rapid rise of mammals could have happened. It was (and remains) a huge, unsolved mystery. Life was not supposed to suddenly disappear, and life was not supposed to suddenly appear. Evolution was a slow and gradual process, requiring millions of years to happen. But the answers were not forthcoming.

This boundary has been named the Cretaceous-Tertiary or K-T extinction event. The *k* stands for the German word for chalk which is from the Latin word for Cretaceous, the last Period of the Mesozoic. The *T* stands for Tertiary, which had been the first Period of the Cenozoic. The Tertiary has been subsequently changed to *Pg*, which stands for Paleogene, now the first system of the Cenozoic. So, now the extinction event is termed, K-Pg or Cretaceous-Paleogene extinction event or boundary.

Era		System	Series
C E N O Z O I C	the last 65 million years right after the great extinction event (K-T)	Quaternary – "Fourth"; 2 million years in length	Holocene – "entirely new"
			Pleistocene – "most new"
		Neogene – "new beginnings"; 23 million years in length	Pliocene – "more new"
			Miocene – "less new"
	Cenozoic means, "recent life" *Age of Mammals*	Paleogene – "ancient beginnings"; 40 million years in length	Oligocene – "few new"
			Eocene – "new dawn"
			Paleocene – "old recent"
The Great K-T or Cretaceous-Tertiary Extinction Event or Boundary			
M E S O Z O I C		**Period**	
		Cretaceous	
		Jurassic	
	Mesozoic means, "middle life" *Age of Dinosaurs*	Triassic	

The extinction boundary that separates the Age of Dinosaurs from the Age of Mammals

Then in the 1970s a new idea about Earth history began to take precedence – Neocatastrophism. This is the idea that Earth history has been shaped by catastrophes. However, they were localized, and they did not change secular geologic time. It's just that the word *catastrophe* was no longer a dirty word, as it had been since the early 1800s. Because of the new catastrophism, in 1980 the catastrophic disappearance of the dinosaurs was suggested. An asteroid hit the Earth 65 million years ago and caused the sudden extinction of the dinosaurs. Scientists enthusiastically received this idea and the public quickly caught on. It was so successful that it has been fully accepted into public life as the scientific reason for why the dinosaurs suddenly disappeared. The idea appeals to scientists because it fits the definition of a naturalistic event. Anyone can visualize an asteroid causing a lot of damage, even a nuclear winter. Movies have been made about it. But then paleontologists began to express doubts. It is clear from the fossils that some modern birds, plants, insects, reptiles, amphibians and mammals existed alongside the dinosaurs and even survived the K-T or K-Pg extinction event. How could a massive global extinction event account for selective extinction? We must take another look at this.

If we apply our Basic Biblical Framework to this problem, it would look more like the following description: There was indeed a great extinction or death event which took place about 4,500 years ago – The Genesis Flood. The Bible tells us in Genesis 7 that all land-dwelling, air-breathing animals outside the ark perished. Many other creatures would have perished also. Billions of marine creatures would have been buried by churned-up sediments. But the Bible is clear, all land-dwelling, air-breathing animals outside the ark perished. The story does not stop there. The floodwaters subsided and the animals inside the ark disembarked, spread out and began to multiply. Because of the tectonic forces released by the Flood, volcanism continued for a while, causing drastic and rapid climate change. This climate change was the catalyst for more death. Many of the animals that were beginning to reproduce would not survive the changes. Dinosaurs would have been drastically affected, as the fossil record indicates that they were probably accustomed to warmer weather. So, a type of extinction has taken place because of the Flood. But it most likely has not been total for many animals. Dinosaurs were seen by ancient people, as the book of Job records at least two beasts, described like dinosaurs, that were actively living with man.

Let's redraw the Geological Time Table to make it fit our Biblical framework and therefore a reinterpretation of the fossil evidence.

The Last 4,500 years – Initial and On-going Effects

4500 years	The Geology	The Climate	The Flora and Fauna
	Initial mountain building, volcanism, earthquakes, local flooding	Initial rapid climate changes; on-going effects	Initial rapid and drastic extinction; variation in those that survived; on-going extinction
	A catastrophic flood, lasting a little over a year (What geologists call the K-T extinction Event		
	Pre-Flood Period lasting 1,600 years		

Initially, volcanism and earthquakes would have continued after the Flood, albeit slowly declining. There would have been initial rapid climate change resulting in more death, and probably extinction, of certain animals. Following the rapid melting of the glacial maximum, due to the decline of volcanism, there would have been localized catastrophic flooding and more rapid climate change to one of a much colder, dryer climate worldwide. Eventually our temperatures and climates of today would have been achieved, but not before many more animals and some humans would have perished as a result.

We now live in an environmentally hostile world where the earth is weak and restless geologically, climate change is relatively unpredictable, and deserts are increasing in size.

The dinosaurs were just some of the many creatures that would have perished in the aftermath of the great Genesis Flood.

Thought Questions

1. What is a bone-bed? What are the characteristics of bone beds? How do bone beds relate to our topic, Dinosaurs and the Flood?

2. Dinosaur Monument is part of a bigger geological formation. What is the name of this formation and how big is it?

3. What is the significance of the lack of plant fossils at Dinosaur National Monument?

4. Explain the problem with the idea of an extinction caused by an asteroid.

5. What is the *death pose*? How does this condition relate to our topic, Dinosaurs and the Flood?

6. Why is the word *extinction* not a good choice of words to describe what has happened to the dinosaurs?

Activity: Go to the web site for Creation Ministries International. In the search box type in the words, disarticulation and bone beds. Read several papers on these subjects and write a report listing the characteristics of each of these subjects.

Lesson Fifteen – Caves and Chemistry
Carlsbad Caverns National Park, and Mammoth Caves National Park

Word challenges: cavern, speleothems, carbonic acid, propionic acid, lactic acid

Carlsbad Caverns National Park

Carlsbad Caverns National Park is located in southeastern New Mexico. It was designated a national park in 1930. The Caverns are primarily noted for Carlsbad Cavern which includes a large cave chamber called The Big Room. It is a natural limestone chamber almost 4,000 feet long, 625 feet wide, and 255 feet high at the highest point. It is the fifth largest chamber in North America and the twenty-eighth largest in the world.

Carlsbad is known for its fantastic limestone dripstone formations called *speleothems*.

The typical secular explanation for the formation of this cave and others like it goes like this. *"An estimated 250 million years ago, the area surrounding Carlsbad Caverns National Park served as the coastline for an inland sea. Present in the sea was a plethora of marine life, whose remains formed a reef. Unlike modern reef growths, the Permian reef contained bryozoans, sponges, and other microorganisms. After the Permian Period, most of the water evaporated and the reef was buried by evaporates and other sediments. Tectonic movement occurred during the late Cenozoic, uplifting the reef above ground. Susceptible to erosion, water sculpted the Guadalupe Mountain region into its present-day state."* National Park Service, Cave Geology: Dissolution and decoration (PDF), retrieved 13 July 2012

It's obvious that this process would have required millions of years to complete. In other words, we could say, small and slow and long ago. But a reading of Genesis 7-8 indicates a geological process that was big and fast and in the recent past. These two views are in conflict with each other.

If we break these explanations down, we have the following key components of cave formation:
- Water
- Calcified marine life
- Receding water
- Burial in sediments
- Uplift
- Erosion
- Time

Both the Basic Biblical Framework and the Uniformitarian framework use these components to explain the presence of and formation of caves. Considering the time limit imposed on us by the Biblical framework, we must ask the following questions:

1. How much water was involved? An "inland sea" does not give me a real solid idea of how much water was involved. A global flood that covered the entire Earth begins to tell me the amount of water that must have initially been involved.
2. The amount of calcium involved in limestone is incomprehensible. It would have taken billions and billions of buried calciferous marine lives to create the amount of limestone involved in the caves around the world. Is it reasonable to think that this limestone was produced by coral reefs as modern geologists insist?
3. Did the water slowly move in and move out? We must know the initial pace of the water, as the huge rooms in caves obviously involved more water in their initial construction than the pace of the dripstones we observe today.
4. Where did all the sediments come from to bury the landscape and billions of marine creatures?
5. How much uplift took place and was it rapid or slow?
6. Was the erosion slow or rapid? If uplift and erosion were slow processes over millions of years, then the Biblical account of the Flood is in error.
7. That brings me to the issue of time – how much time was initially involved in cave formation? Is the time involved in the building of dripstone formations and erosion that we observe today indicative of what has always been?

Caves are quite common in thick beds of marine limestone.

Typically, the analog for the historical formation of caves is what we see going on in them today – *uniformitarianism*. But there are lots of exceptions where this is just simply not the case. The literature is full of examples where animals, birds, and other materials have been covered rapidly by limestone-saturated water within caves. And the slow drip usually pointed out in caves as the pace for cave formation *cannot* explain the formation of the huge rooms that are a part of many caves. The Big Room in Carlsbad Caverns is a prime example. It would have taken a huge amount of water to form The Big Room. The receding floodwaters would have provided the amount and the velocity required to hollow out the huge *caverns* of many caves around the world.

The other factor to consider is that because of the decaying billions of dead things, the acid available to help rapidly erode the freshly laid limestone would have been enormous. This would have greatly affected the rate of erosion.

On-going erosion does occur in caves, but it is not the volume of erosion that would have been active as the floodwaters receded off the Earth, transporting and depositing billions of dead things that were decaying, and producing the acid required to rapidly dissolve limestone. This acid would have produced the huge cavernous rooms that are in most cave systems.

So to summarize how limestone caves (like Carlsbad) could have been formed as a result of the Genesis Flood:
1. The Genesis Flood would have provided the enormous amount of water required to both deposit the limey mud (wet calcium carbonate in a muddy mixture, which hardens into limestone) and sculpt it into caves and the initial massive cave formations.
2. The Genesis Flood as a rapid and catastrophic process would have laid down the limey mud and sculpted it into various cave formations around the world.
3. The enormous amount of limestone present in these caves would have been provided by the calcium carbonate shells of billions of dead invertebrate sea creatures deposited by the Flood.
4. Subsequent erosion of underground limestone would have continued slowly forming some of the speleothems (cave formations, like stalagmites and stalactites) that we see today. However, current dripping of cave water cannot explain the origin of the massive caves. The rate of speleothem formations is highly dependent on the amount of water that continues to drain into these cave systems today. These are influenced by the amount of rain and flooding that an area may receive.

Examples of fossil limestone

Mammoth Cave National Park

Mammoth Cave National Park is located in central Kentucky. It was designated a national park in 1941. The park comprises 133 square miles. The Cave system so far mapped includes 405 miles of passages, and is the world's longest cave system. As typical with most caves, it is composed of Mississippian limestone, which means that it is loaded with marine fossils. The same cave formation principles applied to Carlsbad would also be applied to this cave system.

The huge Rotunda Room at Mammoth Cave National Park

Limestone is highly susceptible to acid. Natural *carbonic acid* in today's environment continues to produce erosion in limestone caves like Mammoth and Carlsbad. Carbonic acid is the result of carbon dioxide mixing with water. It is a weak acid, but does produce erosion in limestone. The all-important question, however, is this. Is the slow production of, and erosion by carbonic acid sufficient to explain the enormous miles of caves and their formations throughout the world?

Let's think about this: When a living thing dies, it immediately begins to decay. This chemical process produces such things as *propionic acid, lactic acid,* methane, hydrogen sulfide, and ammonia. These chemicals rapidly speed up the decay process and would provide the rapid dissolution of limestone in the receding stages of the Flood. This accounts for the large cavernous rooms in limestone caves. Slow dripping of saturated limy water and carbonic acid would not be sufficient for the initial formation of the cave systems around the world.

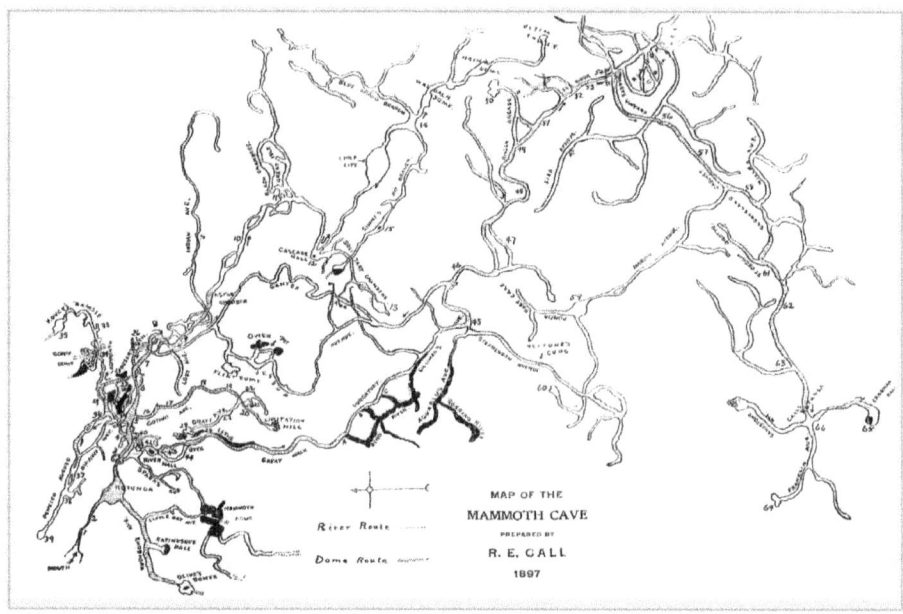

Caves are not only fantastic art displays of nature, they are a laboratory for studying chemistry involving limestone, water, and acid. Caves are also testimonies to the power of a global flood that would have mixed the ingredients to

- produce the limestone,
- destroy the calcareous marine animals
- produce the lime and acid derived from decay

It also would have been the carrier and transporter of it all in the trillions and trillions of gallons of water the Flood would have produced. Initially the water would have been extremely violent in its breaking up and transporting of sediments and dead things. And then as the water started to recede off the Earth, the water would have carved out caverns and cave systems as it moved violently through freshly-laid limestone with the help of the acid from all the dead things.

Types of limestone formations, called speleothems

Thought Questions

1. What is limestone? Why makes it dissolve so easily and quickly?

2. How is the framework of uniformitarianism applied to cave formation?

3. What are the components that would have been provided via the Genesis Flood that would explain cave formation?

Activity: Research the types of dripstone formations found in caves. Propose an explanation of their formation using the Basic Biblical Framework.

Appendix A
What is Deism?

Why spend time expounding on this out-of-date religion? Most people have no idea what it means or how it influenced modern geology. And most people don't care. For most people, science has replaced the outmoded beliefs of religion. Only those who are superstitious or who are fools believe religious books over science.

Deism formed the basis for developing a naturalistic worldview in the 1800s. And that naturalistic worldview gave us uniformitarianism and modern geology. Christian leaders of the 1800s seem to have accepted Deism as more or less a sort of wayward branch of Christianity. One of those branches, the Unitarian Church, had as members some of the most intellectual thinkers of the 1800s. Because of its influence on Western Civilization, it is critical that we understand just what is wrong with Deism and how it conflicts with Biblical Christianity.

Deism comes from a Latin word, *deus* meaning god. Whether it was the intention of the intellectual elite of the 1800s to distinguish this word from the Greek word for god, *theos*, is not known for sure. But it bears thinking about. It is true that the Catholic Church used the Latin language in its services. And that may be why the word *deus* was used. But it is also true that the King James Version of the Bible used the New Testament Greek language and the word for god in Greek is *theos*. Because Deism was a rebellion against the church status quo, it would make sense that the Greek word may have been avoided. Deists distinguished themselves from Theists. Theists were the Protestants. And, sadly, the antagonists to the 19th Century intellectual elitism came from the Protestants, who had gone astray from the Biblical faith. Whatever the reason, the word Deism is not the Christian word for God. That is Theism. The New Testament was in Greek. So, although the Deists believed in a god, it was not the god of the Bible, but some abstract being who was not involved in the affairs of man.

There was no authoritative statement or creed that made Deism an official religion. Deists were varied in their beliefs. Generally, they were not atheists. They seem to have shared skepticism about the supernatural and miracles. They generally did not embrace revelation, i.e., the Scriptures, over man's reason. For the Deist, whatever could not be ascertained by human reason was not genuine science and therefore not worth much. Deists generally thought that the Bible was just another set of myths. Deism, however, advocated a god who probably had created, but they were skeptical about his direct involvement in his creation. God was distant to the Deist and could only be known through human reason. God created natural laws that governed the natural world and created in man a knowledge of those natural laws. Whatever was natural in its cause and effect, was the correct view. Miracles, therefore, were simply man's religious explanations for things he could not explain naturally. According to the deists, our human reason gives us all the information we need.

It is at this point of theism vs. naturalism that Deism diverged from Christianity. It amazes me that there weren't more Christian leaders in the 1800s who elucidated this error and call Deism what it was – a Christian heresy.

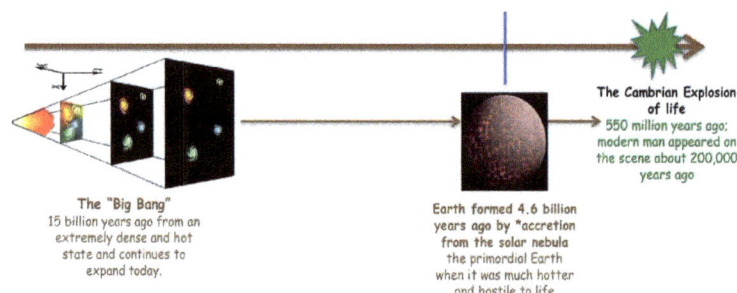

History of the Earth - The Secular View

The "Big Bang" 15 billion years ago from an extremely dense and hot state and continues to expand today.

Earth formed 4.6 billion years ago by *accretion from the solar nebula the primordial Earth when it was much hotter and hostile to life

The Cambrian Explosion of life 550 million years ago; modern man appeared on the scene about 200,000 years ago

Main elements of the Secular View of the History of the Earth:
1. **Naturalistic** - without the aid or intervention of a God
2. **Evolution** - the amazing progression of matter from "something" to intelligent man; simple to complex life
3. **No purpose** for this history other than the amazing way matter has unfolded to produce the beautiful earth and life all in its own power and timing
4. **Volcanoes** produced our early oceans and atmosphere
5. **Oceans were the cradle of life** on earth, then plants on land, then amphibians, reptiles, mammals and finally man after 15 billion years
6. **Man**, rather than being the pinnacle of creation, is a negative force, and is destroying its beautiful earth and systems.

*Accretion is the growth of particles into a massive object by gravitationally attracting more matter, typically gaseous matter

The very foundation of a secular worldview is Deism, which advocates naturalism. Naturalism is at the heart of every aspect of modern geology.

Naturalism is not just an attempt to explain things without reference to a god or a Bible. Today it is an emotional allegiance to human autonomy and reason. It is anathema to a Deist, Atheist and Agnostic to rely in any form on a god who holds us accountable for our every thought and action. Deism was the midpoint between Theism and Atheism. But ultimately it led to atheism in Western Civilization. Naturalistic thinking evolved as the driving force behind modern geology and then evolution.

Naturalism, the child of Deism, is a Christian heresy. It is the opposite of a Biblical view of God. To deny that God has been and continues to be involved in His creation is false. The Scriptures are replete with examples of God's involvement in His creation through prophets, miracles and through His Son, Jesus Christ. This is historical Christianity. This is the reason why I cannot accept explanations for the age of the Earth that claim to be backed by science. It is an oxymoron to claim something is scientific when in fact it deals with issues that cannot be observed, tested or repeated. The age of the Earth is not a scientific subject. It cannot be observed, tested or repeated. Therefore, it is a philosophical and historical issue.

Scriptures that teach God's involvement in His creation include the following:
Psalm 100:5, "For the LORD is good; His lovingkindness is everlasting and His faithfulness to all generations."
Psalm 102:25-26, "Of old You founded the earth, and the heavens are the work of Your hands. Even they will perish, but You endure; and all of them will wear out like a garment; like clothing You will change them and they will be changed."
Psalm 103:7, "He made known His ways to Moses, His acts to the sons of Israel."

Psalm 104:5-9, "He established the earth upon its foundations, so that it will not totter. It is secure. You covered it with the deep as with a garment; the waters were standing above the mountains. At Your rebuke, they fled, at the sound of Your thunder they hurried away. The mountains rose; the valleys sank down to the place which You established for them. You set a boundary that they may not pass over, so that they will not return to cover the earth."

Psalm 104:14, "He causes the grass to grow for the cattle, and vegetation for the labor of man..."

Psalm 104:24, "O LORD, how many are Your works! In wisdom, You have made them all; the earth is full of Your possessions."

Psalm 105:8-11, "He has remembered His covenant forever, the word which He commanded to a thousand generations, the covenant which He made with Abraham, and His oath to Isaac. Then He confirmed it to Jacob for a statute, to Israel as an everlasting covenant, Saying, 'To you I will give the land of Canaan as the portion of your inheritance.'"

Psalm 105:16-17, "And He called for a famine upon the land; He broke the whole staff of bread. He sent a man before them, Joseph, who was sold as a slave."

Psalm 105:26, "He sent Moses His servant, and Aaron, whom He had chosen."

Throughout the history of Israel, God was faithful to His promise He had made to Abraham and continues to fulfill it. These passages of Scripture do not allow for a Deistic God. They teach that He is *theos*. God's name is revealed in Exodus chapter three as, "I AM"; the ever-present one, the one who never changes, and the one who is always faithful to what He has spoken. This is not the Deist concept of God. And therefore, I reject anything that claims to be able to discern the remote history of the Earth through scientific investigation. That is a deception, for scientifically it cannot be done.

Appendix B

An Expanded Explanation of Radiometric Dating

Introduction
Early in the 1800s, part of the Enlightenment involved rejection of the Biblical history that included the Genesis account of Creation and the Global Flood. This was not done based on any scientific discovery, but purely through a change in philosophical thinking. "Enlightened" people would no longer depend on religious stories or myths, like the Bible. And because the Bible taught a recent creation and flood, these "enlightened" thinkers developed ideas that would go in the opposite direction involving hundreds of millions of years of slow geologic development and evolution. Today this thinking has become science doctrine.

Then in 1896 radioactivity was discovered. It seemed to show that certain elements decayed over time – some very quickly and some very slowly, seemingly on the order of hundreds of millions of years. Geologists began to consider this radioactive process as a possible way to objectively and scientifically date the history of the earth. This sort of clock has become known as radiometric dating. It seemed like the final nail in the coffin of Genesis was driven.

What is radiometric dating? Radiometric dating is a system that uses the process and known present rates of radioactivity to project the age of a rock providing certain assumptions are accepted.

What is radioactivity? Radioactivity is the tendency for certain elements, called *isotopes*, to lose energy. Some types of energy are extremely dangerous to living things. The loss of energy from an isotope is measured as a decay per second, called the Becquerel, after its discoverer Henri Becquerel in 1896. Elements that decay are radioactive.

What is an isotope? The word isotope is from the Greek word that means *same place*. An isotope occupies the *same place* on the Periodic Table as its counterpart. For example, the normal carbon atom (C-12, or carbon-12) is found in the 14th Group. Carbon-14 would occupy the same place, but it has something wrong with it.

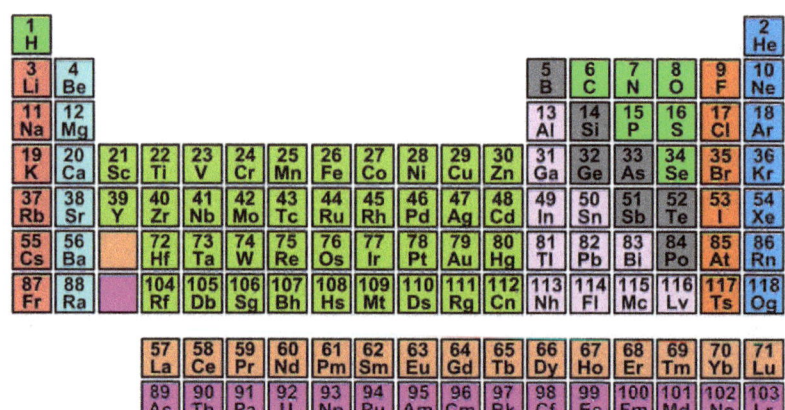

The Periodic Table developed by Dmitri Mendeleev in 1869; only 63 elements were known at the time.

Atomic configuration – Each atom (element) has a special arrangement of atomic particles called protons (+) and neutrons (0) (which form the nucleus) and electrons (-). These are all uniquely and specially configured to give each atom its characteristics. This is by design. The universe is built from atoms of various elements. In the picture below, the number in the upper right-hand corner is the atomic number. It tells us how many protons and electrons an atom has. In the example of carbon-12 – six protons and six electrons. The number of electrons always equals the number of protons. The letter in the center is the chemical symbol for the element – in this example, carbon. The number on the bottom is the atomic mass, which equals the combined number of protons, and neutrons. So, how many neutrons does a normal carbon atom have? The answer is 6. In carbon-14, however, the atomic mass is 14, which means that it has two more neutrons than normal carbon and therefore is unstable. It decays. It loses energy. It is radioactive.

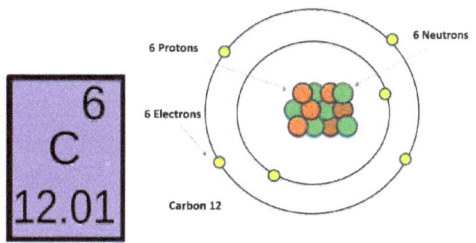

Now, here is an important question. Why, if after the 6th day of creation when God pronounced everything "very good", do we have abnormal atoms from the original atoms that God created? No one has ever been able to scientifically show why we have abnormal atoms. Human beings have almost all of the naturally occurring atoms within their structure in just the right amounts – even arsenic (As)! It might very well be that radioactivity was a vital part of the inner workings of the earth but nonexistent on the surface of the earth at creation. At the point the fountains of the great deep burst open, radioactivity could have been released in great amounts producing the false readings that are so abundant today.

What happens to a radioactive element? Because an abnormal element has an abnormal number of atomic particles, it tends to "shed" those atomic particles with the net effect of changing into a stable element. Because these atomic particles can be identified and observed, a statistical prediction can be made. If carbon-14 loses particles at so many Becquerels or decays per second, then theoretically it should turn into another stable element, given certain assumptions and enough time. We will discuss these assumptions later. In the example below in the process of radioactivity, an atomic particle is released. There are several different kinds of these particles depending on the type of radioactivity. The releasing of this particle is energy released in the form of Becquerels and these can be measured as they are observed.

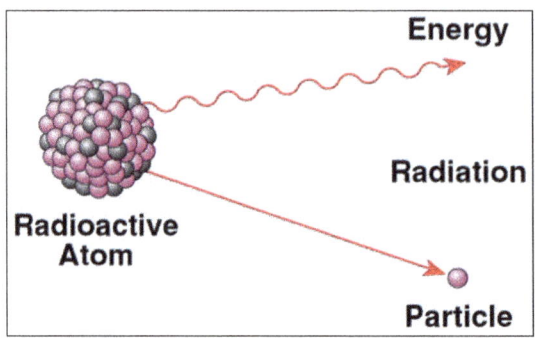

This picture is a theoretical representation.

Radioactivity

A common explanation for radioactivity: Radioactivity is caused when an atom, for whatever reason, seeks to lose some of its energy, in the form of atomic particles. This is called decay. (Different kinds of particles can be released.) It does this because it wants to shift from an unstable configuration to a more stable configuration. The energy that is released when the atom makes this shift is known as radioactivity. In other words, radioactivity is the act in which an atom releases radiation suddenly and spontaneously. Some radioactivity is extremely dangerous. Some is not. The important thing to remember is that radioactivity can be measured. No one has ever been able to definitively say why radioactive elements exist.

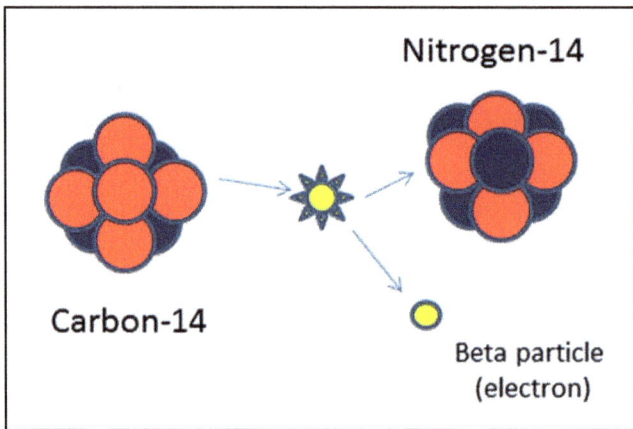

This diagram shows what, theoretically, happens when radioactive Carbon–14 decays, and in so doing, becomes stable Nitrogen–14, over a long period of time.

Radioactive decay

If radioactive elements lose particles in the decay process, what happens to the original radioactive element? Obviously, there is going to be a change that takes place. The radioactive element theoretically becomes something else. I say theoretically, because the decay process can require an immense amount of time to complete and would not be able to be observed throughout the entire process. The unstable element that is decaying is called the *parent* and the stable element into which it theoretically eventually changes is called the *daughter*. Many of the radioactive elements used in radiometric dating require huge amounts of time and no one has recorded or witnessed the entire decay process. For example, the time it takes for ½ of Carbon-14 to decay, measured at present measurable rates is a little over 5,000 years. Then in another 5,000 years ½ of what is left decays and so on until it is all gone. Has anyone ever observed the complete process to know whether it has actually completed this process or not? How could they? Below is an illustration of the idea of half-life. Theoretically all that is left at the end of the decay process is the stable daughter element.

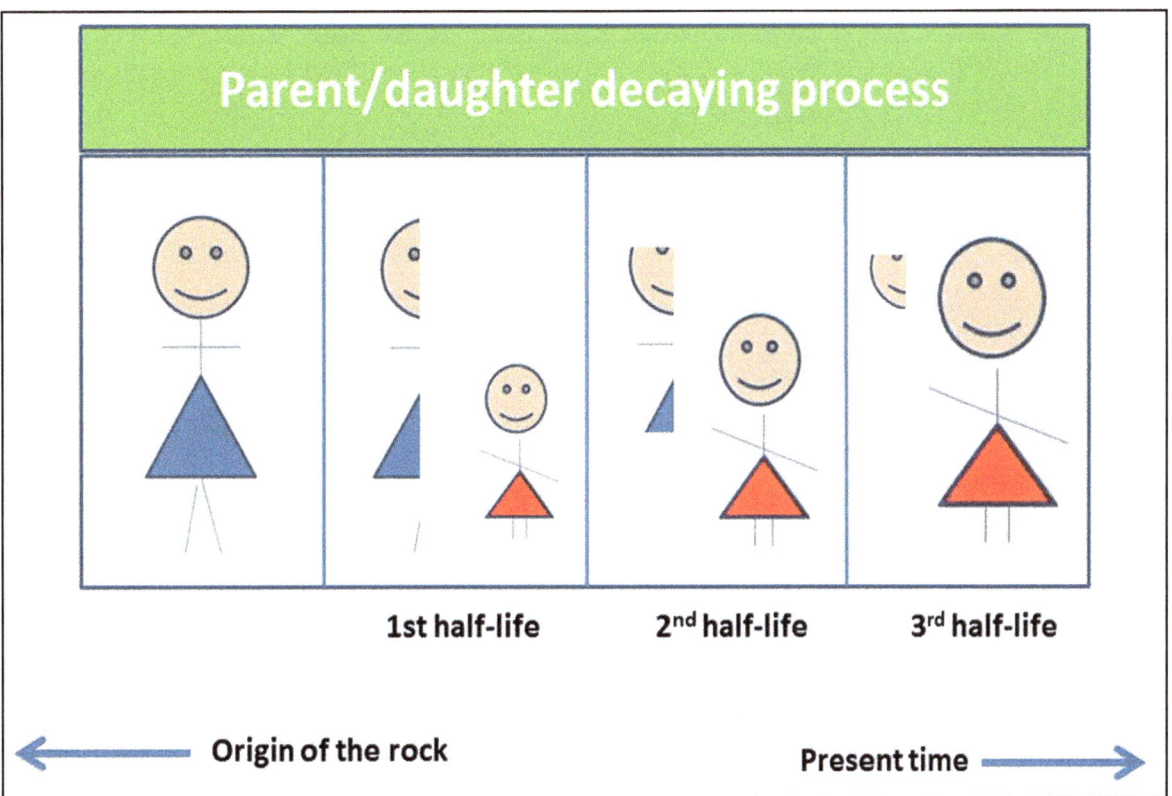

The half-life of the radioactive element is determined by the present Becquerel. The time is then projected forward or backward to arrive at a time when the process theoretically started or when it would theoretically end. In our Carbon-14 atom above, the stable element, into which C-14 would decay, is Nitrogen-14. This is a chemical prediction based on what we know of present radioactive decay. Remember that the process, if totally accurate, would take thousands of years to complete. No one has been around that long.

Uranium decay
Radioactive uranium238 (U-238) is a very heavy element. But notice its half-life from the picture below! So, the question is, has anyone ever observed this? The answer is obviously "No". Then how do we know that it operates this way? We can *guess* based on the chemistry of radioactive decay. It is a matter of statistics. If U-238 were to continue to lose particles like it seems to do in the present, then over such and such a length of time it will go through a "chain" of decays until a stable element is arrived at – in this case, Lead-206.

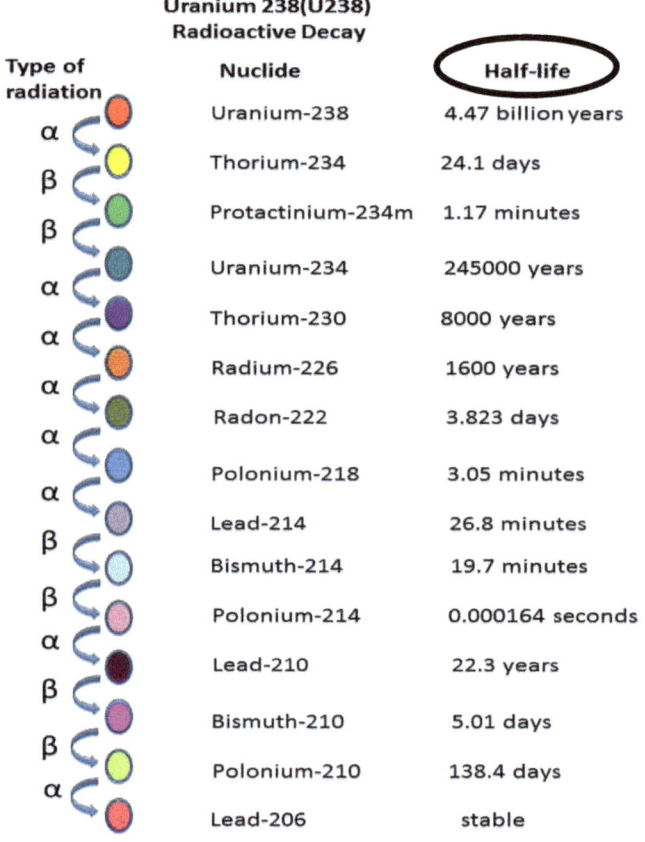

U-238-Lead-206 is a commonly used method in calculating the radioactive decay/age of a rock. But if no one was around to actually test or record the entire process from beginning to end, then how do we know this process works? We don't! Geologists make *assumptions* based on the present observation of radioactivity. Does that sound familiar? It should. It is an application of the philosophy known as uniformitarianism – *the present is the key to the past.* So, by assuming the present has always been true in the past, geologists can use the process of radioactive decay to date rocks. How do they do this?

Only rocks that are thought to have been molten at one time are considered to be valid specimens. So, fossils cannot be directly dated because of contamination. Fossils are dated by another means, also by way of uniformitarianism. That is, creatures change in the present. Therefore, creatures have changed in the past and given enough time, have changed radically into other types of creatures – evolution. Look at the Geologic Column (next page). It is presented as a scientific fact. But in reality, it is a hypothetical idea. The column in its entirety has not been found anywhere on the Earth. Some of the rock layers do occur in an apparent order, but there is also an explanation for this apparent order that modern geologists reject. It is a general order (with many exceptions) produced by the sorting action of the Genesis Flood. If the fountains of the great deep burst open first, as Genesis states, then one would expect that marine (sea) creatures would have been the first victims of the Flood. As the Earth became inundated with water, other creatures would have followed as the Flood swallowed them up. Man, being the most mobile, would have most likely been buried last, if at all. Creatures swimming in order to survive would have ultimately drowned and their remains would have been eaten by fish or sharks. The apparent order in the rock layers has been assumed by geologists to be the result of an evolutionary process, having rejected the idea of a global flood.

The Geologic Column or Time Scale or Time Table
As the idea of change in creatures expanded to an evolutionary view where one type of creature was thought to have changed into an entirely different creature, so the amount of time for nature to accomplish this was also necessarily expanded from several thousand years to 550 million years by the end of the 19th Century. The basic Geologic Column as we have today was set by the mid to late 19th Century. In other words, the concept of time plus evolution was a philosophical shift from a Divine Creation taught in the Book of Genesis to a totally naturalistic and atheistic one taught through uniformitarianism. Today many people think that the Geologic Column and the millions of years it involves is a scientific fact supported by radiometric dating. Remember that radioactivity was not even discovered until after the Geologic Column had been in place for many years. Most people are unaware of this. It was not science that formulated the Geologic Column, but philosophy. It was a naturalistic attempt to explain the apparent order of fossils. That is, marine fossils on the bottom of layers, dinosaurs closer to the top. It is an idea, and an accepted idea, but not proven science.

EON	ERA	PERIOD	MILLIONS OF YEARS AGO
Phanerozoic	Cenozoic	Quaternary	1.6
		Tertiary	66
	Mesozoic	Cretaceous	138
		Jurassic	205
		Triassic	240
	Paleozoic	Permian	290
		Pennsylvanian	330
		Mississippian	360
		Devonian	410
		Silurian	435
		Ordovician	500
		Cambrian	540
Proterozoic	Late Proterozoic Middle Proterozoic Early Proterozoic	Ediacaran 635-543 MYA	2500
Archean	Late Archean Middle Archean Early Archean		3800?
Pre-Archean			

Extinction Event 65 Million Years Ago (End of Cretaceous)

Extinction Event (End of Permian)

Cambrian - 540 Million Years

How is a radiometric date arrived at?
Let's start with a volcanic rock. And let's say it is chemically analyzed to contain 80 parts of uranium and 20 parts of lead. Is the rock young or old? Since more of the uranium is present than lead, then the rock must be young. If the ratio was reversed, then the opposite would be true – *given the following assumptions*:
1. We *assume* that the initial state of the rock started with a certain amount of uranium and no lead. In other words, the initial state of the rock is *assumed*.
2. We *assume* that there was no lead present at the start of the process.
3. We *assume* that no uranium came from some other source.
4. We *assume* that all of the lead that is present came from the decay of uranium and that it did not come from some other source.

5. We *assume* that the decay rate or process has not been interfered with from some other means or sources.

So long as we go with these *assumptions*, we can statistically figure an age for the rock.

But what happens when different ages are obtained, as in the case of Mount St. Helens lava, which was known to be ten years old, but which dated in the hundreds of thousands of years. Again, geologists simply *assume*:

1. You or the laboratory contaminated the samples.
2. You or the laboratory made a mistake in calculations.

If radiometrically obtained ages do not fit with geologists' assumptions about past behavior of radioactivity, they are thrown out.

But what if the sample dates different ages using different dating methods, as in the case of rocks from the Grand Canyon where different dates for the same rock were obtained using different methods. Then the "umpire" is the Geological Time Scale (Geologic Time Table). Since dates to 550 million years old were worked out in the 1800s, and have already been agreed upon, if the date for the rock appears to be too old or too young, then the dates are either thrown out or selected. The date which is the closest to what the geologist thinks it is becomes the accepted date. *And that's radiometric dating.*

Radiometric assumptions are not reliable

What might have caused radioactivity to change in the past? The Genesis Flood would have added a tremendous amount of heat to rocks. It would have also added a tremendous amount of hot water to the rocks. And in fact, it has been shown that radioactive decay was rapidly sped up in the past in several substances:

1. Carbon-14 was discovered in diamonds thought to be at least several billion years old. According to present measurements of the decay of C-14, there should be absolutely no detectable C-14 after around 100,000 years.
2. Carbon-14 was discovered in petrified wood thought to be at a few hundred million years old.
3. Recent observed lava flows in New Zealand and in Hawaii dated excessively old according to modern radiometric dating methods.
4. In addition, if radiometric dates need to be checked by another source, the Geologic Time Table, for accuracy and reliability, then are they reliable methods?

These examples could have been affected by Earth's processes given the proposition that the Genesis Flood was a real, historical event. If this is so, then thousands if not hundreds of millions of years' worth of radioactive decay occurred in a matter of a few months at some time in Earth's past. The radioactive process has been dynamically affected. This means that the radiometric clocks that geologists have developed are really not reliable. Even though radioactivity is an observable fact, the measurement and predictions of this process have been affected by something else and should not be used as a basis for age/dating. It would be like hanging on to a clock or watch that ticks but is always gaining or losing time. Why depend on it? It's a clock, but is it reliable?

Why do geologists continue to use an unreliable method?
Geologists continue to use the radiometric dating methods despite the obvious flaws. Why? Although there is rarely an open public admission of these errors, the radiometric dating methods provide the so-called "scientific" alternative to the young Earth conclusions of the Bible. At the center of the radiometric dating rational are the influence of The Enlightenment and consequently the extremely narrow view of uniformitarianism.

Mount St. Helens recent dacite dome

A recent lava flow in Hawaii

How long does this radioactive process take to complete its transformation? That really depends on the size of the atom. Carbon-14 takes less time than the much larger atom of uranium-238. **Half-life** is the term used to describe the time required, probabilistically, for half of the unstable, radioactive atoms in a sample to undergo radioactive decay. Based on the Becquerels per second (decays per second), it would theoretically take around 5,730 years for half of a carbon-14 atom to decay. Then after another 5,730 years, theoretically another half-life will have occurred, and so on until carbon-14 is gone and nitrogen-14 is complete. (Try this exercise: draw a circle and divide it in half. That would equal 5,730 years. Then divide the remainder in half, and so on. How many years would it theoretically take for the carbon-14 atom to be totally changed?) In another example, the half-life of uranium-238 is 4.468 billion years!

Has this ever been observed? How could it? The creation itself is only about 6,000 years old! In addition, no one has ever been around long enough to have recorded it. Scientists simply assume that given the observable decays per second and enough time, such and such would happen. It must be remembered here that **radioactivity** (the tendency for isotopes to lose energy and consequently atomic particles) is a fact of science. **Radiometric dating** is not a scientific fact but an application of the scientific fact of radioactivity. Radiometric dating is based on the philosophy of uniformitarianism. As it applies to radioactivity, uniformitarianism would state that what we observe in the present has always been going on in the past at the same rate and in the same way we observe now. And therefore, given enough time, an isotope would transform into a stable element. This sounds scientific and reasonable. However, if we introduce a historical global flood into the picture, then uniformitarianism is a false premise. Also, a global flood would have changed or interrupted normal decay rates and even introduced contaminates into otherwise natural processes.

Appendix C

Final Exam – Multiple Choice

1. The most abundant kind of lava produced by calderas is
a) Basalt
b) Rhyolite
c) Andesite
d) Dacite

2. A pyroclastic rock that is formed from ash and small rocks is
a) Tuff
b) Ash
c) Shale
d) Tuffa

3. The first national park in the world was
a) Crater Lake National Park
b) Petrified Forest National Park
c) Yosemite National Park
d) Yellowstone National Park

4. The philosophy that states, "The present is the key to the past"
a) The Enlightenment
b) Nebular Hypothesis
c) Uniformitarianism
d) Secularism

5. A crack in the Earth is called
a) A split
b) An earthquake
c) An eruption
d) A rift

6. The word tectonics has to do with
a) Tearing Earth apart
b) Building Earth together
c) Repairing Earth
d) Removing Earth

7. Deism is the belief that
a) God is actively engaged in His creation
b) God is concerned about His creation
c) God is uninvolved in His creation
d) God hates His creation

8. Cross-bedding is
a) Alternating currents left in sandstone
b) Alternating rifts left in sandstone
c) Alternating fossil directions left in sandstone
d) Alternating layers left in sandstone

9. A laccolith is
a) An empty crater
b) A raised mount
c) A sandstone structure
d) A would-be volcano

10. Scoria is
a) A type of andesite
b) A type of dacite
c) A type of basalt
d) A type of tuff

11. Disarticulation means
a) Disconnected bones
b) Disconnected tissue
c) Unburied bones
d) Nonfossilized bones

12. The secular extinction boundary for the dinosaurs is called
a) The Q-T extinction event
b) The K-T extinction event
c) The P-G extinction event
d) The C-Pg extinction event

13. Mount St. Helens is a
a) Caldera
b) Shield volcano
c) Cinder Cone
d) Stratovolcano

14. Rhyolite has a high amount of
a) Quartz
b) Iron
c) Sodium
d) Olivine

15. Andesite is produced mostly by
a) Shield volcanoes
b) Spatter cones
c) Calderas
d) Stratovolcanoes

16. Underwater landslides are called
a) Cross-beds
b) Turbidity currents
c) Swift currents
d) Seamounts

17. Another word for speleothem is
a) Limestone
b) Stalagmite
c) Dripstone
d) Calcification

18. The modern Rock Cycle divides rocks into
a) two main rock types
b) four main rock types
c) three main rock types
d) five main rock types

19. The Cascade Range of Mountains has been called
a) The most weathered set of mountains
b) The smallest set of mountains
c) The most colorful set of mountains
d) The most complex set of mountains

20. The Teton Mountains are a type of
a) Volcano block
b) Fault block
c) Rift block
d) Canyon block

21. The most thermally active volcano in the Cascade Range is
a) Mt. Baker
b) Mt. Adams
c) Mount St. Helens
d) Mt. Rainier

22. Cave limestone is most often made of
a) Fossil vertebrates
b) Fossil dinosaurs
c) Fossil marine creatures
d) Fossil plants

23. The ropy kind of lava is called
a) a'a
b) Pahoehoe
c) Scoria
d) Cinder

24. A volcano that emits showers of hot lava is called
a) Stratovolcano
b) Spitting cone
c) Cinder cone
d) Basalt cone

25. A black form of rhyolite lava is called
a) Tachylite
b) Bentonite
c) Obsidian
d) Tuff

Appendix D

Rock-forming Minerals

Throughout this book, many rocks and minerals have been mentioned. There are thousands of minerals in the world, but the vast majority of rocks include and are made of just these twelve minerals. These twelve include six light-colored, and six dark-colored minerals. If you learn these, rock identification is much easier. For a complete study in rock identification, see my book, Rock Identification Made Easy.

Appendix E

Answers to the Exam

1. b
2. a
3. d
4. c
5. d
6. b
7. c
8. a
9. d
10. c
11. a
12. b
13. d
14. a
15. d
16. b
17. c
18. c
19. d
20. b
21. a
22. c
23. b
24. c
25. c

Index of Words From Word Challenges

Analog 156, 168
Andesite 56, 59, 63, 108, 141, 150,
Archipelago 100
Ash 13, 19, 21, 28, 31, 32, 35, 41, 51, 52, 53, 56, 58, 61, 62, 63, 66, 84, 85, 86, 87, 93, 147, 149, 151, 152, 153, 154, 160, 185
assumption (assumes) 7, 8, 56, 59, 60, 62, 87, 93, 101, 143, 176, 177, 180, 181, 182, 184
attrition 37
basalt 5, 21, 35, 36, 56, 59, 63, 77, 78, 83, 85, 86, 88, 89, 91, 96, 101, 102, 103, 105, 107, 110, 111, 112, 113, 116, 128, 130, 141, 142, 152,
basement rocks 23, 123, 127, 128, 133, 142
batholith 65
bone beds 158, 161, 165
breccia 41, 136, 141
canyon 18, 28, 75, 79, 82, 109, 124, 125, 126,
carbonic acid 170
cast 90
catastrophism (catastrophic) 13, 15, 133, 144, 156, 160, 161, 162
cavern 168, 170
chronology 9, 11
cinder cone 77, 84, 85, 89, 95, 99, 100, 107, 108, 110, 113
clast, clastic 35, 126, 127, 136, 145,
clay 28, 51, 52, 55,
climate change 66, 117, 118, 161, 163
composite volcano 56
conglomerate 28, 29, 30, 31, 141, 158, 160
cross-bedding 75, 76
cyanobacteria 119, 120, 121
dacite 63, 64, 147, 151, 183
death pose 159, 165
deposition 30, 32, 51, 53, 113, 132
diorite 65, 68, 136, 143
disarticulation 158, 159, 161
ecological zone 54
erosion 8, 28, 33, 51, 52, 53, 74, 76, 79, 80, 81, 103, 109, 132, 144, 166, 167, 168, 170
erratic boulders 66, 67
extinction 161, 162, 163, 181
fault-block mountain 70
fossil 27, 28, 29, 31, 32, 33, 34, 35, 37, 38, 40, 41, 42, 43, 44, 45, 46, 47, 48, 49, 51, 52, 53, 54, 71, 72, 119, 120, 126, 127, 130, 131, 132, 145, 153, 158, 159, 160, 161, 162, 163, 169, 180, 181
framework 7, 8, 9, 14, 16, 19, 25, 28, 41, 47, 53, 60, 66, 67, 72, 81, 87, 93, 94, 101, 102, 109, 113, 118, 119, 124, 125, 126, 127, 128, 129, 130, 133, 142, 144, 153, 155, 158, 163, 167
fumarole 15, 113, 140
gabbro 143
geology 7, 8, 9, 14, 25, 28, 60, 61, 65, 66, 67, 74, 94, 109, 123, 124, 125, 132, 144, 145, 154, 160
glacial troughs 66

glacier 13, 23, 24, 57, 58, 60, 65, 66, 67, 70, 77, 84, 93, 117, 118, 119, 120, 136, 139, 140, 141, 149, 150
global warming 117, 118
gneiss 23, 66, 136, 143, 145
granite 23, 65, 66, 67, 68, 71, 128, 136, 143, 144
granodiorite 68, 143
graveyard fossils 161
hoodoo 79
hornitos 113
hydrological 133
igneous 128, 144, 145
isotope 59, 176, 184
laccolith 81, 82
lactic acid 170
lahar 40, 41, 57, 58, 148, 156
lava tube 92, 103, 104, 113
lungfish 32
maars 113
massif 136, 142
mesa 79
metamorphic rock (metamorphism) 66, 70, 71, 72, 126, 127, 133, 136, 142, 143, 144, 145
micro-vertebrates 51, 52
naturalistic (naturalism) 7, 8, 16, 19, 124, 162, 173, 174, 181
obsidian 91, 92
orogeny 70, 72
pahoehoe 101, 102, 107, 110, 112, 113,
petrified (petrification) 13, 24, 27, 28, 29, 30, 31, 32, 33, 76, 182
photosynthesis 120
pit crater 113
planation 27, 28, 74, 109
plutonic rock 65, 66, 68, 72, 126, 127, 128, 133, 136, 142, 143, 145
propionic acid 170
pumice 85, 86, 87, 113
pyroclastic 35, 56, 57, 59, 62, 63, 85, 86, 87, 147, 149, 151
quarry 38, 158, 159
quartz 21, 28, 56, 62, 63, 68, 75, 86, 87, 88, 91, 126, 128, 130, 143, 151,
quartzite 143, 145
radiometric dates 41, 60, 93, 101, 128, 182
rift zone 93
sandstone 28, 30, 35, 51, 52, 71, 72, 74, 75, 76, 77, 79, 80, 81, 82, 126, 127, 143, 158, 160
schist 66, 127, 143
science 6, 7, 43, 59, 60, 71, 101, 109, 124, 125, 126, 173, 174, 176, 181, 184
scoria 85, 86, 110, 111
scute 51
secular 6, 7, 8, 9, 15, 16, 19, 20, 28, 40, 41, 43, 46, 47, 52, 60, 61, 62, 71, 76, 84, 93, 94, 101, 113, 118, 124, 125, 130, 132, 144, 145, 158, 161, 162, 166, 174,

sedimentary 24, 32, 35, 40, 41, 71, 75, 81, 118, 119, 126, 127, 130, 132, 133, 136, 141, 144, 145, 159, 160, 161
serpentinite 143
shale 41, 118, 119,
shield volcano 84, 85, 86, 88, 91, 96, 97, 108, 113, 115,
silica 21, 28, 29, 41, 75, 160,
silicic 35
siltstone 41, 51, 52, 127
speleothems 113, 166, 168, 171
stratovolcano 40, 56, 59, 77, 84, 85, 139, 141, 147, 148, 150,
stromatolites 119. 120
supernaturalism 125
tectonic 19, 66, 70, 72, 126, 163, 166
tephra 35
tuff 20, 21, 59, 62, 63, 85, 86, 108,
turbidity current 76
unconformity 132
uniformitarianism 7, 125, 144, 168, 173, 180, 181, 183, 184
varve 47
vertebrates 38
viscous (viscosity) 63, 88, 102, 128
volcanic vent 98, 110, 111
worldview 7, 16, 43, 101, 125, 127, 128, 173, 174

Credits

Lesson One

A Basic Biblical Framework: Patrick J. Nurre, 9; The Genesis Flood, The Order of Events and Geological Implications: Patrick Nurre, Copyright information: **1st picture**: http://astrogeology.usgs.gov/search/details/ganymede/geology/ganymede_sim3237_database/zip; **2nd picture**: URL: http://geomaps.wr.usgs.gov/parks/noca/nocageol4c.html; **3rd picture**: public domain; **4th picture**: URL: http://geomaps.wr.usgs.gov/parks/province/rockymtn.html; **5th picture**: USGS Photo, 11;

Lesson Two

Lewis and Clark Expedition map: By Victor van Werkhooven - Own work, This file was derived from: Carte Lewis-Clark Expedition-en.png, Public Domain, https://commons.wikimedia.org/w/index.php?curid=32915075, 15; Buffalo: Photo by Vicki Nurre, 15; Hot spring: Photo by Vicki Nurre, 16; Fumarole: Photo by Karen Dennis, used by permission, 16; Mud pot: Photo by Vicki Nurre, 16;Geyser: Photo by Vicki Nurre. 16; Earthquake map: *NPS map. Earthquake data provided courtesy University of Utah*, 17; Black Dragon's Caldron (2): Photos by Patrick Nurre, 18; Attachment point, Earthquake Lake: Photo by Patrick Nurre, 19; Hebgen Lake earthquake damage: Photo courtesy of the USGS, 20; Yellowstone caldera: *USGS Volcano Map of Yellowstone National Park*, 21; Yellowstone eruptions: By National Park Service. - US NPS., Public Domain, https://en.wikipedia.org/w/index.php?curid=39165505, 21; Sheepeater Cliff: Photo by Vicki Nurre, 22; Lavas and tuff (3): Photos by Patrick Nurre, 22; Geyserite (2) Photos by Vicki Nurre, 22; Riverside Geyser: Photo by Patrick Nurre, 23; Travertine (2): Photos by Patrick Nurre, 23; Travertine in mammoth Hot Springs: Photo by Patrick Nurre. 23; Beartooth Mountains: Photo by Vicki Nurre. 24; The Gallatin Range: By Mike Cline - Own work, CC BY-SA 4.0, https://commons.wikimedia.org/w/index.php?curid=38185091. 24; Glacial cirques (2): Photos by Patrick Nurre, 25; Cobbles: Photo by Patrick Nurre, 25; The Absaroka Range:By National Park Service - http://www.nps.gov/yell/slidefile/scenics/outsidepark/Images/09175.jpg, Public Domain, https://commons.wikimedia.org/w/index.php?curid=532167, 25;

Lesson Three

Map of Petrified Forest National Park: By National Park Service; converted from PDF to PNG format by User:Finetooth using GNU Image Manipulation Program (GIMP) software - official park map, Public Domain, https://commons.wikimedia.org/w/index.php?curid=11907421, 27; Colorado Plateau Map: Public Domain, https://commons.wikimedia.org/w/index.php?curid=65130, 28; Petrified log covered with conglomerate: Photo by Patrick Nurre, 29; Close-up of conglomerate: Photo by Patrick Nurre, 29; Sandstone formation: Photo by Patrick Nurre, 30; Rounded stones: Photo by Patrick Nurre, 30; Petrified rock on top of conglomerate: Photo by Patrick Nurre, 31; Volcanic ash deposits: By Finetooth - Own work, CC BY-SA 3.0, https://commons.wikimedia.org/w/index.php?curid=11931890, 31; Phytosaur fossil: Photo courtesy of the National Park Service, 32; Reptiles: By Richie Diesterheft from Chicago, IL, USA - Uploaded by FunkMonk, CC BY 2.0, https://commons.wikimedia.org/w/index.php?curid=7690475, 32; Coelophysis: Photo courtesy of the National Park Service, 32; Petrified trees: Photo by Patrick Nurre, 33; Planed surface: Photo by Patrick Nurre, 33; The Hagerman Fossil Beds National Monument: Photo by Patrick Nurre, 35; Red Rock Pass: By Wilson44691 - Own work, Public Domain, https://commons.wikimedia.org/w/index.php?curid=10941096, 36; Path of the Lake Bonneville Flood: Photo by Patrick Nurre, 36; Tumbled basalt boulder: Photo by Patrick Nurre, 36. Post ice age floods map: By Fallschirmjäger - Own work.Based on this map by Laura DeGrey, Myles Miller and Paul Link of Idaho State University, Dept. of Geosciences.Inset North America map from File:America-blank-map-01.svgLakes in Western Great Basin sourced from USGS, Extent of Pleistocene Lakes in the Western Great Basin by Marith Reheis, 1999.Further sourced from SERC, Carleton College, Reheis and Bright, 2009.Coastline based on The Pacific Northwest Coast: Living With the Shores of Washington and Oregon, Paul D. Komar and USGS map.iThe source code of this SVG is valid., CC BY-SA 3.0, https://commons.wikimedia.org/w/index.php?curid=25056130, 37; The Hagerman Horse: By NPS photo - http://www.nps.gov/hafo, Public Domain, https://commons.wikimedia.org/w/index.php?curid=2611067, 37; John Day Fossil Beds National Monument Map: By National Park Service; converted from PDF to PNG format by User:Finetooth using GNU Image Manipulation Program (GIMP) software - Harpers Ferry Center, Public Domain, https://commons.wikimedia.org/w/index.php?curid=16077638, 40; Oregon environment painting: Mural by Larry Felder, courtesy of the National Park Service, 40; Black Butte, Oregon: Photo by Aselfcallednowhere - Own work, CC BY-SA 3.0, https://commons.wikimedia.org/w/index.php?curid=11217893, 41; Brontothere: By Nobu Tamura (http://spinops.blogspot.com) - Own work, CC BY 3.0, https://commons.wikimedia.org/w/index.php?curid=19459933, 41; Strata of the John Day Fossil Beds National Monument: By Finetooth - Own work, CC BY-SA 3.0, https://commons.wikimedia.org/w/index.php?curid=16539464, 42; Brontothere: Photo by Postdlf from w, CC BY-SA 3.0, https://commons.wikimedia.org/w/index.php?curid=2655565 Entelodont: Photo by Peteron. - http://www.copyrightexpired.com/earlyimage/bones/display_osborn_dinohyus.htm, Public Domain, https://commons.wikimedia.org/w/index.php?curid=5817273, 42; Hippo-like animals and hyena-like animals Hippo-like animals: Phot by I, Sailko, CC BY-SA 3.0, https://commons.wikimedia.org/w/index.php?curid=20393583. Hyena-like animals: By Apokryltaros at English Wikipedia, CC BY-SA 3.0, https://commons.wikimedia.org/w/index.php?curid=6532814, 42; Tiger-like animal: Photo by Gally242 - Own work, CC BY-SA 3.0, https://commons.wikimedia.org/w/index.php?curid=18333278, 43; Oreodont: Photo by Patrick Nurre, 43; John Day environments (2): Photos courtesy of the National Park Service, 43,44; Fossil fish: Photo by Vicki Nurre, 46; Fossil Butte National Monument: Photo by Leaflet - Own work, CC BY-SA 4.0, https://commons.wikimedia.org/w/index.php?curid=45159446, 46; Insect and shrimp fossils (10): Photos by Patrick Nurre, 47-48; Fossil fish: Photo courtesy of Tim and Candey (Earths.Ancient.Gifts.com).48; Fossil leaf: Photo courtesy of Tim and Candey (Earths.Ancient.Gifts.com).48; Fossil palm leaves, water plants and fossil fish: Photo by Vicki Nurre, 48; Fossil fish: Photo by Patrick Nurre. 49; Spires and pinnacles, Badlands National Park: Photo by Carol M. Highsmith - This image is available from the United States Library of Congress's Prints and Photographs division under the digital ID highsm.04594. Public Domain, https://commons.wikimedia.org/w/index.php?curid=41344760, 51; Small mammal vertebrae: Photo by Patrick Nurre, 52; Turtle Shell: Photo by Patrick Nurre, 52; Ammonite: Photo by Patrick Nurre, 52; Dinosaur bones: Photo by Patrick Nurre, 52; Mammal coprolites: Photo by Patrick Nurre, 52; Large mammal vertebra: Photo by Patrick Nurre, 52; Teeth from a camel: Photo by Patrick Nurre, 53;

Fossils of the White River Badlands: Photo by Patrick Nurre, 53; Badlands, eastern Wyoming: Photo by Wilson44691 Photograph taken by Mark A. Wilson (Department of Geology, The College of Wooster). [1] - Own work, Public Domain, https://commons.wikimedia.org/w/index.php?curid=4328394, 53; Badlands, southwestern South Dakota: Photo by Ronincmc - Own work, CC BY-SA 4.0, https://commons.wikimedia.org/w/index.php?curid=46944662, 53; Badlands, eastern Montana: Photo by Ankyman at the English language Wikipedia, CC BY-SA 3.0, https://commons.wikimedia.org/w/index.php?curid=3600454, 53. Fossil Titanothere: Photo by Patrick Nurre, 54;

Lesson Four

Mt. Rainier National Park: Photo by Samuel Kerr - Own work, CC BY-SA 3.0, https://commons.wikimedia.org/w/index.php?curid=21013358, 56; Map of Mt. Rainier: Courtesy of the USGS, 57; Map, lahar hazards: By © Sémhur / Wikimedia Commons, Public Domain, https://commons.wikimedia.org/w/index.php?curid=3207214, 58; Tacoma,WA: Photo by Lyn Topinka (USGS) - http://vulcan.wr.usgs.gov/Imgs/Jpg/Rainier/Images/Rainier84_mount_rainier_and_tacoma_08-20-84.jpg [1], Public Domain, https://commons.wikimedia.org/w/index.php?curid=337686, 58; Andesite lava: Photo courtesy of the National Park Service, 59; Andesite lava: Photo by Patrick Nurre, 59; Rhyolite lava: Photo by Patrick Nurre, 59; Tuff: Photo by Patrick Nurre, 59; Mt. Lassen: Photo is Public Domain, https://commons.wikimedia.org/wiki/File:Lassen_Peak_Manzanita_Lake.jpg, 61; Mud pot: Photo by SkepticalRaptor - Own work, CC BY-SA 3.0, https://commons.wikimedia.org/w/index.php?curid=19149845, 61; Map, Lassen Volcanic National Park: Photo courtesy of the USGS, 62; Mt. Lassen before and after 1914: Photo courtesy of the National Park Service, by LassenNPS - Hot Rock and Lassen Peak eruption, CC BY 2.0, https://commons.wikimedia.org/w/index.php?curid=44303992, 62; Rhyolite lava: Photo by Patrick Nurre, 63; Tuff: Photo by Patrick Nurre, 63; Dacite lava: Photo by Patrick Nurre, 63; Andesite lava: Photo by Patrick Nurre, 63;

Lesson Five

Yosemite National Park: Photo by Diliff - Own work, CC BY-SA 3.0, https://commons.wikimedia.org/w/index.php?curid=26413820, 65; Half-Dome: Photo by Jon Sullivan - [1], Public Domain, https://commons.wikimedia.org/w/index.php?curid=3643796, 66; Striated and rounded granite: Photo by Patrick Nurre, 66; Glacial erratic boulders: Photo by Patrick Nurre, 67; Glacial trough in Yosemite: Photo by Patrick Nurre, 67; Louis Agassiz: Photo by John Adams Whipple - http://www.picturehistory.com/product/id/3700, Public Domain, https://commons.wikimedia.org/w/index.php?curid=16445240, 68; Granite: Photo by Patrick Nurre, 68; Granodiorite: Photo by Patrick Nurre, 68; Diorite: Photo by Patrick Nurre, 68;

Lesson Six

The Tetons: Photo by Jon Sullivan, PD Photo. - PD Photo, http://pdphoto.org/PictureDetail.php?mat=pdef&pg=8145, Public Domain, https://commons.wikimedia.org/w/index.php?curid=3537847, 70; The Teton Fault Block: By NPS image from http://www2.nature.nps.gov/geology/parks/grte/, Public Domain, https://commons.wikimedia.org/w/index.php?curid=5712991, 71; Mt. Moran: Photo by Acroterion - Own work, CC BY-SA 3.0, https://commons.wikimedia.org/w/index.php?curid=11608275, 72;

Lesson Seven

Photos: Sandstone (2): Photos by Heidi Noggle, 74; Sandstone: Photo by Patrick Nurre, 74; Sandstone in Zion National Park: Photo by Patrick Nurre, 74; Planation surface: Photo by Patrick Nurre, 74; Zion Canyon: Photo by Diliff - taken by Diliff, CC BY-SA 3.0, https://commons.wikimedia.org/w/index.php?curid=305224, 75; Diagram of idealized cross-bedding formation: By Nwhit - Own work, CC BY-SA 3.0, https://commons.wikimedia.org/w/index.php?curid=17594578, 75; Cross-bedded formation: Photo by Patrick Nurre, 76; By Brian W. Schaller - Own work, FAL, https://commons.wikimedia.org/w/index.php?curid=30556200, 77; Preserved water ripples: Photo by Patrick Nurre, 77; Large water-tumbled basalt boulders: Photo by Patrick Nurre, 78; Map of Capitol Reef National Park: By National Park Service, Harpers Ferry Center - http://www.nps.gov, Public Domain, https://commons.wikimedia.org/w/index.php?curid=1087131, 78; Bryce Canyon amphitheater: Photo by Jean-Christophe BENOIST - Own work, CC BY 3.0, https://commons.wikimedia.org/w/index.php?curid=11617169, 79; Canyonlands National Park: Photo by Phil Armitage - http://www.philarmitage.net/canyonlands/canyon4.html, Public Domain, https://commons.wikimedia.org/w/index.php?curid=813234, 80; Delicate Arch: Photo by Palacemusic - Prise de vue personnelle, CC BY-SA 3.0, https://commons.wikimedia.org/w/index.php?curid=6020905, 80; Magmatic rock of the La Sal Mountains (2): Photos by Patrick Nurre, 81; Diagram of a laccolith buried: By en:User:Erimus, User:Stannered - en:Image:Laccolith.JPG, Public Domain, https://commons.wikimedia.org/w/index.php?curid=3713849, 81; Round Mountain: Photo by Patrick Nurre, 82;

Lesson Eight

Crater Lake: Photo by Epmatsw - Own work, CC BY-SA 3.0, https://commons.wikimedia.org/w/index.php?curid=32942620, 84; The Pinnacles near Crater Lake: Photo by No machine-readable author provided. Llywrch assumed (based on copyright claims). - No machine-readable source provided. Own work assumed (based on copyright claims)., CC BY 2.5, https://commons.wikimedia.org/w/index.php?curid=329786, 85; Relief map of Crater Lake: By USGS - [1], Public Domain, https://commons.wikimedia.org/w/index.php?curid=1149851, 85; Tuff: Photo by Patrick Nurre, 86; Pumice: Photo by Vicki Nurre, 86; Vesicular basalt: Photo by Patrick Nurre, 86; Ash tuff: Photo by Patrick Nurre, 86; Scoria: Photo by Patrick Nurre, 86; Volcanic Explosivity Index: Vicki Nurre, 87; Caldera eruption: By U.S. GEOLOGICAL SURVEY and the NATIONAL PARK SERVICE - http://pubs.usgs.gov/fs/2002/fs092-02/ (saved as PNG), Public Domain, https://commons.wikimedia.org/w/index.php?curid=667211, 88; Crater Lake: Photo by Zainubrazvi - Own work, CC BY-SA 3.0, https://commons.wikimedia.org/w/index.php?curid=2033872, 89; Map of the Newberry Volcano/Caldera National Monument: Courtesy of the USGS, 90; Newberry National Volcanic Monument: Photo by L. Moclock (talk · contribs) - Own work by the original uploader, CC0, https://commons.wikimedia.org/w/index.php?curid=24426388, 91; Newberry Crater: Photo by USGS photo by Lyn Topinka - http://vulcan.wr.usgs.gov/Imgs/Jpg/Newberry/Images/Newberry85_newberry_caldera_obsidian_flow_08-20-85.jpg, Public Domain, https://commons.wikimedia.org/w/index.php?curid=1988402, 91; Obsidian lava flow: Photo courtesy of the USGS, 91; Obsidian: Photo by Ji-Elle - Own work, CC BY-SA 3.0, https://commons.wikimedia.org/w/index.php?curid=15527635., 92; Obsidian (2): Photos by Patrick Nurre, 92; Lava River Cave: Photo by Dave Bunnell / Under Earth Images - Own work, CC BY-SA 4.0, https://commons.wikimedia.org/w/index.php?curid=48892566, 92; The Brothers Fault Zone: Public Domain, https://en.wikipedia.org/w/index.php?curid=22025810, 93; Rifts and faults: By J. Johnson - Own work, CC BY-SA 3.0, https://commons.wikimedia.org/w/index.php?curid=11836456, 94;

Lesson Nine

Hawaii map: By NordNordWest - own work, usingUnited States National Imagery and Mapping Agency dataU.S. Geological Survey (USGS) data, CC BY-SA 3.0 de, https://commons.wikimedia.org/w/index.php?curid=7015666, 96; The Big Island, map: By Mapmaunaloa.png: Hawaii Volcano Observatory, USGSderivative work: Richardprins (talk) - Mapmaunaloa.png, Public Domain, https://commons.wikimedia.org/w/index.php?curid=10852262, 97; Mauna Kea: Photo by Patrick Nurre, 97; Mauna Loa map: Courtesy of the USGS, 98 Kilauea active crater: Photo courtesy of the USGS, 98; Satellite image of the Pu'u'O'o volcanic field: Picture courtesy of the USGS, 99; Volcanic cinder cone of Pu'u'O'o: Photo by G.E. Ulrich, USGS. Cropping by Hike395 (talk · contribs) - USGS, Public Domain, https://commons.wikimedia.org/w/index.php?curid=4412830, 99; Aerial view of Pu'u'O'o: By J.D. Griggs - USGS HVO, Public Domain, https://commons.wikimedia.org/w/index.php?curid=925189, 99; Lava from the Pu'u'O'o *cinder cone*: Photo by Madereugeneandrew - Own work, CC BY-SA 4.0, https://commons.wikimedia.org/w/index.php?curid=39413714, 100; Pahoa and Pu'u'O'o, outlined by burnt vegetation: Photo by Madereugeneandrew - Own work, CC BY-SA 4.0, https://commons.wikimedia.org/w/index.php?curid=39415332, 100; The Hawaiian Island Chain of volcanoes: Photo by Jacques Descloitres - File:Hawaje.jpgOriginal source: NASA. Image courtesy Jacques Descloitres, MODIS Land Rapid Response Team at NASA GSFC. (IotD Date: 2003-06-03. IotD ID: 15304), Public Domain, https://commons.wikimedia.org/w/index.php?curid=11359528, 100; Hawaiian Islands chain: By United States Geological Survey (USGS) - http://hvo.wr.usgs.gov/volcanoes/, Public Domain, https://commons.wikimedia.org/w/index.php?curid=196173, 101; Vesicular basalt: Photo by Patrick Nurre, 102; *Pahoehoe* lava: Photo by Patrick Nurre, 102; Basalt with olivine: Photo by Patrick Nurre, 102; *A'a* basalt lava flow: Photo by Patrick Nurre, 102; The volcanic mountains of Oahu: Photo by Patrick Nurre, 103; Water-tumbled basalt boulders: Photo by Patrick Nurre, 103; Lava tube: Photo by Vicki Nurre, 104; Map of the Craters of the Moon National Monument: By NPS map, Original uploader was Mav at en.wikipedia - Transferred from en.wikipedia, original source: NPS image from http://www.nps.gov/crmo/pphtml/maps.html, Public Domain, https://commons.wikimedia.org/w/index.php?curid=6308196, 105; Map of Idaho and Craters of the Moon National Monument: By NPS image from http://www.nps.gov/crmo/pphtml/maps.html, Public Domain, https://commons.wikimedia.org/w/index.php?curid=23352488, 106; Tree mold: By James Tolbert - Own work, CC BY-SA 3.0, https://commons.wikimedia.org/w/index.php?curid=28408528, 106; 'A'a lava field: Photo by Vicki Nurre, 106; Cinder cone: Photo by Vicki Nurre, 107; Basalt cinders: Photo by Vicki Nurre, 107; Pahoehoe flow: Photo by Vicki Nurre, 107; Raton-Clayton Volcanic Fiele: Photo by Eric T Gunther - Own work, CC BY 3.0, https://commons.wikimedia.org/w/index.php?curid=38515393, 108; Capulin Volcano: Photo by R.D. Miller, USGS - U.S. Geological Survey Photographic Library, Public Domain, https://commons.wikimedia.org/w/index.php?curid=8266839, 108; Map, Rio Grande Rift: Courtesy of the National Park Service, 109; View around Capulin: Photo by Leaflet - Own work, Public Domain, https://commons.wikimedia.org/w/index.php?curid=7488984, 109; Zuni-Bandera Volcanic Field: Photo by Lee Siebert, Smithsonian Institution - http://www.volcano.si.edu/world/volcano.cfm?vnum=1210-02-&volpage=photos&photo=033087, Public Domain, https://commons.wikimedia.org/w/index.php?curid=6663242, 110; Map, Colorado Plateau:Rangerdavid, own work, October 2012, under the Creative Commons Attribution-Share Alike 3.0 Unported license, 110; Basalt ribbon bombs (2): Photos by Patrick Nurre, 111; Scoria: Photo by Patrick Nurre, 111; Aphanitic basalt: Photo by Patrick Nurre, 111; Columnar vesicular basalt: Photo by Patrick Nurre, 112; Pahoehoe lava flow: Photo courtesy of the USGS, 112; Oxidized pahoehoe lava: Photo by Patrick Nurre, 112; Lava Beds National Monument: Photo by Beej Jorgensen en:User:Beej71 - en-Wikipedia, CC BY-SA 2.0, https://commons.wikimedia.org/w/index.php?curid=1609886, 113; Oregon section of The Volcanic Legacy Scenic Byway: By U.S. Department of Transportation, Federal Highway AdministrationThe original uploader was Howcheng at English Wikipedia - http://byways.org/fws/byways/2587 (for latest version), Public Domain, https://commons.wikimedia.org/w/index.php?curid=26178217, 114; Map of the California section of The Volcanic Legacy Scenic Byway: By (Original text: "National Scenic Byways Program (byways.org)Original image found at http://www.byways.org/library/display/25851/California_Section.gifUpdated image athttp://www.byways.org/explore/byways/2587/travel.html?map=California_Sectionhttp://library.byways.org/a/static_maps/000/000/583/California_Section.gif), Public Domain, https://commons.wikimedia.org/w/index.php?curid=8394553, 115; Medicine Lake Volcano: Photo by Daniel Mayer en:User:Maveric149 - en-Wikipedia, CC BY-SA 3.0, https://commons.wikimedia.org/w/index.php?curid=1609905, 115;

Lesson Ten

Hidden Lake and Bearhat Mountain: By NPS Photo - Official Website: http://www.nps.gov/archive/glac/gallery/072600b.htm, Public Domain, https://commons.wikimedia.org/w/index.php?curid=6139829, 117; Shale with ripple patterns: Photo by Patrick Nurre, 118; Ripple marks in the shale of the Belt Supergroup: Photo by Vicki Nurre, 118; Glacier National Park: By Ken Thomas (talk · contribs) - Own work by the original uploader, Public Domain, https://commons.wikimedia.org/w/index.php?curid=1566506, 119; Fossil stromatolites: Photo by James St. John - Stromatolites (Snowslip Formation, Belt Supergroup, Mesoproterozoic, 1.44 Ga; Glacier National Park, Montana, USA) 3, CC BY 2.0, https://commons.wikimedia.org/w/index.php?curid=35287361, 120; Photosynthesis: By Daniel Mayer (mav) - original imageVector version by Yerpo - Own work, GFDL, https://commons.wikimedia.org/w/index.php?curid=20722530, 120; The chloroplast: By Kelvinsong - Own work, CC BY 3.0, https://commons.wikimedia.org/w/index.php?curid=26247252, 121; Cyanobacteria:By Matthewjparker - Own work, CC BY-SA 3.0, https://commons.wikimedia.org/w/index.php?curid=24119069, 121;

Lesson Eleven

Grand Canyon: Photo by Akarsh Simha - Own work, CC BY-SA 3.0, https://commons.wikimedia.org/w/index.php?curid=23336216, 123; Grand Canyon: Photo by Patrick Nurre, 126; Sandstone (2): Photos by Heidi Noggle, used by permission, 127; Sandstone: Photo by Patrick Nurre, 127; Non-fossil limestone: Photo by Heidi Noggle, used by permission, 127; Fossil limestone: Photo by Patrick Nurre, 127; Schist: Photo by Patrick Nurre, 127; Granite: Photo by Patrick Nurre, 128; National Geographic magazine: Photo by Vicki Nurre, 129; The San Francisco Volcanic Field: By USGS - http://wrgis.wr.usgs.gov/fact-sheet/fs024-02/images/fig5_large.jpg - United States Geological Survey, Public Domain, https://commons.wikimedia.org/w/index.php?curid=1263629, 129; Basalt: Photo by Patrick Nurre, 130; Fossil bryozoa: Photo by Patrick Nurre, 130; Fossil clam: Photo by Patrick Nurre, 130; Fossil squid: Photo by Patrick Nurre, 130; Gastropods (2): Photos by Patrick Nurre, 130, 131; Trilobite: Photo by Patrick Nurre, 131; Fossil coral: Photo by Patrick Nurre, 131; Fossil brachiopods: Photo by Patrick Nurre, 131; Fossil Crinoids: Photo by Patrick Nurre, 131; Worm borrows: Photo by Vicki Nurre, 131; Fossil tracks: Photo courtesy of National Park Service, 132; Fossil leaves: Photo courtesy of National Park Service, 132; Grand Canyon Rock Layers: By Patrick Nurre, 132; Colorado River: Photo by Chensiyuan - Own work, GFDL, https://commons.wikimedia.org/w/index.php?curid=9756217, 133;

Lesson Twelve

North Cascade Range: Photo by G310ScottS – Own work, CC BY-SA 3.0, https://commons.wikimedia.org/w/index.php?curid=16756162, 136; Map of the North Cascades National Park: Courtesy of the National Park Service, 137; Pelton Peak: Photo By Daniel Hershman – Flickr, CC BY 2.0, https://commons.wikimedia.org/w/index.php?curid=2752656, 138; The Pickett Range: Photo by Walter Siegmund – Own work, CC BY 2.5, https://commons.wikimedia.org/w/index.php?curid=2795370, 138; Glacier Peak: Photo by Walter Siegmund – Own work, CC BY 2.5, https://commons.wikimedia.org/w/index.php?curid=593451, 139; Mt. Baker: Photo by Ken McGee, USGS - http://volcanoes.usgs.gov/vsc/images/image_mngr/1000-1099/img1042.jpg, Public Domain, https://commons.wikimedia.org/w/index.php?curid=10412842, 139; Sampling fumarole gas at Sherman Crater: Photo by W. Chadwick – Archived United States Geological Survey link, Public Domain, https://commons.wikimedia.org/w/index.php?curid=3491081, 140; Map of Mt. Baker: Public Domain, https://en.wikipedia.org/w/index.php?curid=17344181, 140; Volcanic breccia: Photo by Vicki Nurre, 141; Andesite: Photo by Vicki Nurre, 141; Rhyolite: Photo by Vicki Nurre, 141; Basalt in columnar form: Photo by Patrick Nurre, 142; Mt. Shuksan: Photo by Frank Kovalchek from Anchorage, Alaska, USA – Mt. Shuksan reflected in a small tarn on the Artist Point trail Uploaded by hike395, CC BY 2.0, https://commons.wikimedia.org/w/index.php?curid=9744866, 142; Diorite : Photo by Patrick Nurre, 143; Gabbro: Photo by Patrick Nurre, 143; Granodiorite: Photo by Vicki Nurre, 143; Schist: Photo by Patrick Nurre, 143; Quartzite: Photo by Patrick Nurre, 143; Gneiss: Photo by Vicki Nurre, 143; Serpentinite: Photo by Vicki Nurre, 143; The Rock Cycle: By Woudloper/Woodwalker - I photoshopped this, Public Domain, https://commons.wikimedia.org/w/index.php?curid=1707248, 144;

Lesson Thirteen

Mt. St. Helens: Photo: by Patrick Nurre, 147; 360⁰ panorama photo of Mt. St. Helens crater rim: Photo by Gregg M. Erickson (talk · contribs) - Own work, CC BY 3.0, https://commons.wikimedia.org/w/index.php?curid=9651194, 147; Map of the Cascade Range: By Shannon - Background and river course data from http://www2.demis.nl/mapserver/mapper.asp, mountain elevation/location data from Wikipedia, GFDL, https://commons.wikimedia.org/w/index.php?curid=9774848, 148; Mt. St. Helens pre-1980 eruption: Photo by Harry Glicken, USGS/CVO - USGS photo of Mt. Saint Helens, Public Domain, https://commons.wikimedia.org/w/index.php?curid=634392, 149; Mt. St. Helens post-eruption: Photo by Harry Glicken - USGS Cascades Volcano Observatory, Public Domain, https://commons.wikimedia.org/w/index.php?curid=672199, 149; Toppled forest: Photo courtesy of the USGS, 149; Stripped trees: Photo courtesy of the USGS, 149; Crater Glacier: Phot by Bergman Photographic Services (under contract to U.S. Geological Survey) - Schilling, Steve P.; Paul E. Carrara, Ren A. Thompson, and Eugene Y. Iwatsubo (2004). "Posteruption glacier development within the crater of Mount St. Helens, Washington, USA". Quaternary Research 61 (3): 325–329., Public Domain, https://commons.wikimedia.org/w/index.php?curid=1982478, 150; Promonent points of interest, Mt. St. Helens: Picture courtesy of USGS, 150; Two views of Mt. St. Helens: Photos courtesy of the USGS, 150; Andesite: Photo courtesy of USGS, 151; Dacite: Photo courtesy of USGS, 151; Ash from Mt. St. Helens: Photo by Patrick Nurre, 151 Ash fall in Yakima, WA (2): Photos courtesy of the USGS, 152; Plant life at Mt. ST. Helens: Photo courtesy of the USGS, 152; Mt. St. Helens, 2005: Public Domain, https://commons.wikimedia.org/w/index.php?curid=735753, 153; Lamar Valley: Photo by Mike Cline - Own work, Public Domain, https://commons.wikimedia.org/w/index.php?curid=8876633, 153; Specimen Ridge 1890 and copy or original drawing: By J. P. Iddings - Downloaded from USGS Photo Library: http://libraryphoto.cr.usgs.gov/htmllib/batch82/batch82j/batch82z/batch82/ijp00325.jpg, Public Domain, https://commons.wikimedia.org/w/index.php?curid=8876732, 154; Tree at Specimen Ridge: Photo by John Meyer, used by permission, 154; The log mat: Photo by Stephan Schulz - Own work, CC BY-SA 3.0, https://commons.wikimedia.org/w/index.php?curid=20997835, 155; The Toutle River Badlands: Photo courtesy of the USGS, 155; The Toutle River Badlands: Photo courtesy of the USGS, 156; The Toutle River Badlands: Photo by U.S. government - http://vulcan.wr.usgs.gov/Volcanoes/MSH/SlideSet/ljt_slideset_old.html, Public Domain, https://commons.wikimedia.org/w/index.php?curid=633431, 156;

Lesson Fourteen

Green River Canyon: Photo by CC BY-SA 3.0, https://commons.wikimedia.org/w/index.php?curid=753535, 158; Dinosaur quarry: Photo by Transferred from en.wikipedia to Commons.; originally from http://www.cr.nps.gov/museum/treasures/html/Q/h020.html, Public Domain, https://commons.wikimedia.org/w/index.php?curid=3970352, 159; Death pose: Photo by James St. John - Camarasaurus lentus sauropod dinosaur (Morrison Formation, Upper Jurassic; Carnegie Quarry, Dinosaur National Monument, northeastern Utah, USA) 1, CC BY 2.0, https://commons.wikimedia.org/w/index.php?curid=35841021, 159;
Death pose: Photo by Sebastian Bergmann from Siegburg, Germany - [1], CC BY-SA 2.0, https://commons.wikimedia.org/w/index.php?curid=4295088, 159; A fossilized leg bone: Photo courtesy of the National Park Service, 160; Dinosaur National Monument (2): Photos by Patrick Nurre, 160; The extinction boundary, chart: By Patrick Nurre, 162; The last 4,500 years, chart: By Patrick Nurre, 163;

Lesson Fifteen

"Carlsbad Cavern" Photo by Eric Guinther, User:Marshman - Uploaded to the English Wikipedia August 2003 (deleted page), CC BY-SA 3.0, https://commons.wikimedia.org/w/index.php?curid=64449, 166; Thick marine limestone: Photo courtesy of the National Park Service, 167; Cave in thick marine limestone: Photo courtesy of National Park Service, 167; Fossil limestone: Photo by Manishwiki15 – Own work, CC BY-SA 3.0, https://commons.wikimedia.org/w/index.php?curid=32776278, 169; Fossil limestone (2): Photo by Jim Stuby (talk) – I (Jim Stuby (talk)) created this work entirely by myself.Transferred from en.wikipedia, Public Domain, https://commons.wikimedia.org/w/index.php?curid=17780464, 169; Fossil limestone: Photo by Vicki Nurre, 169; Fossil limestone: Photo by Vicki Nurre, 169; Fossil limestone: Photo by Vicki Nurre, 169; Rotunda Room at Mammoth Cave National Park: Public Domain, https://commons.wikimedia.org/w/index.php?curid=261447, 169; The Mammoth Cave System: By Image extracted from page 153 of The Mammoth Cave of Kentucky. An illustrated manual, etc ..., by HOVEY, Horace Carter – and CALL (Richard Ellsworth). Original held and digitized by the British Library. Copied from Flickr.Note: The colours, contrast and appearance of these illustrations are unlikely to be true to life. They are derived from scanned images that have been enhanced for machine interpretation and have been altered from their originals.This file is from the Mechanical Curator collection, a set of over one million images scanned from out-of-copyright books and released to Flickr Commons by the British Library.View image on Flickr View all images from book View catalogue entry for book.English | Français | +/-, Public Domain, https://commons.wikimedia.org/w/index.php?curid=32129952, 170; Speleothems: Photo by Dave Bunnell / Under Earth Images - Own work, CC BY-SA 2.5, https://commons.wikimedia.org/w/index.php?curid=22613190, 171;

Appendix A

Secular view of world history, chart: Photos Public Domain, chart by Patrick Nurre, 174;

Appendix B

The Periodic Table: Openclipart.com, 176; Carbon atom: Image by Vicki Nurre, 177; Radioactive decay: Source: NRC at http://www.nrc.gov/reading-rm/basic-ref/glossary/full-text.html, 178; Radioactive decay: Image by Vicki Nurre, 178;. Parent-daughter decay: by Vicki Nurre,179; Decay chains: Image by Vicki Nurre, 180; Geologic Time Table: http://www.bing.com/images/search?pq=geologic+time+&sc=8-14&sp=2&sk=IA1&q=geologic+time+scale+chart&qft=+filterui:licenseL2_L3&FORM=R5IR42#view=detail&id=0548D46FA426B2733D713B5A16EE55BFDB889547&selectedIndex=0, 181; Mt. St. Helens Dome: Public Domain. USGS. Found at http://mythphile.hubpages.com/hub/volcano-glossary, 183; Lava flow: Photo by Vicki Nurre, 183;

Appendix D

Rock-forming minerals: Photo by Vicki Nurre, 190;

Patrick Nurre has been a rock hound since childhood and has an extensive rock, mineral and fossil collection, having collected from all over the United States. In 2005, he started Northwest Treasures, which is devoted to designing geology kits for schools. He conducts numerous field trips each year in Washington State to such places as the Olympic Peninsula, Mt. Rainier, Mount St. Helens, the Channeled Scablands, Mt. Baker and Whidbey Island. In addition, he also gives field trips to the volcano loop of Oregon and California, Mt. Hood volcanic area (Oregon), the eastern badlands of Montana and Yellowstone National Park. He is a popular speaker at homeschool conventions, schools, and churches. Patrick pastored in churches for over 40 years. He is the author of sixteen geology textbooks and is a Certified Biblical Creation Ministry Professional.

If you would like to contact Patrick about speaking or field trips: northwestexpedition@msn.com

For a list of speaking topics: NorthwestRockAndFossil.com

www.ingramcontent.com/pod-product-compliance
Lightning Source LLC
Chambersburg PA
CBHW061142010526
44118CB00026B/2845